Cooperating Through Chaos

You only get one life.

You must choose:
Are you going to spend it
stepping on your fellow humans
for your own gain
or
working with them
towards a greater good?

If it's the latter,
this book was written for you.

ANIMAL CENTRIC BOOKS

First published in Australia in 2024 by Animal Centric Books

1 3 5 7 9 10 8 6 4 2

NATIONAL LIBRARY OF AUSTRALIA

A catalogue record for this book is available from the National Library of Australia

Hardback ISBN 978-0-6459803-1-8
Trade Paperback ISBN 978-0-6459803-0-1
Ebook ISBN 978-0-6459803-2-5
Audio ISBN 978-0-6459803-3-2

Animal Centric Books
An imprint of
Animal Centric
Brisbane, Australia

www.animal-centric.com

Cooperating Through Chaos

WHY HUMANS (AND OTHER ANIMALS) WORK TOGETHER

Jill Fernandes, PhD

ANIMAL CENTRIC BOOKS

To Keyoren, whom I was lucky enough to marry.
You're my best friend,
my biggest supporter,
and an inexhaustible source of animal behaviour.

And to my nieces, Ruby and Lily.
May you grow up full of hope.

CONTENTS

AUTHOR'S NOTE

Dates

Please note that in this book, all dates occurring before the year 0 in the Gregorian calendar are expressed in millions of years ago (mya), thousands of years ago, or hundreds of years Before the Common Era (B.C.E.). This is the standard used in astronomy, geology, and paleontology.

Units

The International System of Units (the modern metric system) is used throughout this book, as it is the standard used in scientific literature. English equivalents can be found using:

https://www.unitconverters.net/

Introduction

WHY THIS BOOK?

What the world needs now is cooperation.

Do we need economic reform? Of course. Participatory governance? Definitely. Freedom of expression? For sure. But at the core of all human progress lies our ability to work together. Or, put simply, to cooperate.

I believe that at this point in human history, cooperation isn't just a nice-to-have; it's a must-have. Actually, cooperation has always been a must-have for human beings, which is why our species evolved to work together, rather than apart, in so many different contexts.

But now, in particular—when large segments of society seem to have fallen under the spell of individualism, when we are tempted to think that a combination of military prowess, economic progress, and technological innovation will solve all the world's problems if we just give it time—we need to pull our heads out of the sand and look at the natural history of our species.

If we examine the science of our evolution, we will rediscover an important truth: Humans, like many other animals, have an innate capacity to work together. We cooperate with each other not because it's "nice." We do it to survive.

This book explores the science of human cooperation, taking a journey through millions of years of evolution and drawing on multiple disciplines of science to answer a central question:

What is the glue that binds humanity,
and how can we use it for good?

WHO IS THIS FOR?

I wrote this book for people like me—people who are fascinated with animals, nature, and humanity. I am a scientist by disposition and formal training, but I believe you shouldn't have to do a PhD to engage in scientific inquiry. Humans are naturally inquisitive. To me, anyone who relishes asking questions and building on the understanding of others is a scientist at heart. So, this book is for knowledge seekers (scientists) of all ages and levels of training.

A disclaimer for academic readers: Although science is its backbone, this book is not meant to be the authoritative scholarly review on the subject of cooperation. Robert Axelrod and Lisa D'Ambrosio have put together a wonderful online compendium of studies on the evolution of cooperation, the link to which can be found at the end of Chapter 1. For anyone wanting to go deeper into the subject of cooperation after reading this book, I suggest you start there. You can also look into the Further

Reading and References at the end of each chapter. By the same token, if you have a relevant publication that you think I may have missed, please send it to me via the contact info at the back of this book. I'm a lifelong learner, and I'll be delighted to receive a new treasure.

HOW TO READ THIS BOOK

Sailing

This book was born at the start of the COVID-19 pandemic: March 2020. As the first cases of local transmission appeared in Australia, hysteria swept through the streets of Brisbane. Looking to escape it, my husband and I drove 1,100 km north to the town of Airlie Beach. From there we chartered a sailboat and escaped to the tropical waters surrounding the Whitsunday Islands. We were only able to afford this week-long rental because the charter company had halved their normal prices due to last-minute COVID cancellations. We were lifelong travelers but amateur sailors. We were in over our heads, sailing on our own. But we felt we had to get away or get sucked into pandemonium.

Our week on this sailboat, exposed to the natural elements and away from a society in chaos, unlocked a new door inside my brain. I began to consider the power of the environment in shaping the behavior of humans and other animals. My thoughts leapt to Charles Darwin's theory of natural selection. I began to wonder, *How has humanity been shaped by our environment over the last 6 million years to make us what we are today?*

Because of how eye opening that one week aboard a sailboat in the Whitsundays was to me, every chapter of this book starts with a glimpse into that week. In the 3 years that followed, I sank deeper and deeper into the research along this line of inquiry. No subject matter was out of bounds: I perused theories and studies from evolutionary biology, archaeology, paleoclimatology, neuroscience, game theory, astrophysics, psychology, sociology, and animal behavior. What I discovered in the process is that nature has made humans surprisingly cooperative. I felt compelled to synthesize this knowledge and share it with the world. The product is this book.

Animals

In addition to drawing on lessons from nature that have been revealed to me through sailing, I draw on lessons from other animals. To me, animals are the wisdom of nature, embodied. My life's work is aimed at shaping a better world for all animals. I do this by bridging the neuroscience of how animals feel with the many real-world contexts in which people care for animals. It's an exciting time to be immersed in animal welfare science. Society is starting to ask more and more questions of other animals and actually listen to their responses, rather than making assumptions about what they know or are capable of. When it comes to cooperation, I believe we are only just scratching the surface of everything that animals can teach us. In each chapter, I present a story from elsewhere in the animal kingdom that sheds light on our own cooperative behavior.

Scenes

I've never been drawn much to novels, having always preferred to read nonfiction (my husband likes to joke about how much it shows). But I love movies. So, this book is structured like a movie, in a series of 50 scenes. Each scene is bite sized—you should be able to read one in about 4 to 8 minutes. All the scenes actually happened, but the dialogue has been recreated based on my memory and the memories of others; it's not a word-for-word transcription. This book is meant to be read from start to finish, like a movie is meant to be watched. That being said, if you put the book down for several days and go and do something else, you will not be lost when you pick it up again. The important messages are layered in throughout the book. So, don't worry if you're a person who struggles to concentrate or reads several books at once. My hope is that you'll enjoy this book and learn something interesting, no matter how much or little you read at one time.

WHAT YOU WILL TAKE AWAY

If you read this book to the end, you will have a better understanding of why many animals, including humans, cooperate to survive. Importantly, you will also have a clear roadmap for getting large groups of humans to cooperate so that we can address the urgent issues facing the world today. I want you to understand cooperation, but ultimately, I want you to practice it. Humanity needs you to practice it.

My intention is that you'll walk away from this book with the confidence that we are in this together. Humans are social animals. None of us is here on Earth alone. We are not single threads stretched along the individual timelines of our lives. You cannot pluck one of us from the middle without disturbing the rest. We are interwoven. We belong to each other. And because we belong to each other, we work together through our times of greatest challenge.

We cooperate through chaos.

CHAPTER 1

Stayin' Alive: Survival

I.

I awoke to a crash on the hull. *What the ...* I sat up and froze. On the other side of the fiberglass was a confusion of rain, wind, and waves. Above my head, the deck creaked to a tempo as the boat rocked to and fro.

A stomp on the deck just above me. My neck instinctively crouched. Another stomp. And another. Like the footsteps of a mammoth man, plodding towards the cockpit. *Is someone on board?* I held my breath and awaited more. But amid the cacophony, the only signal I could discern was the repeated creaking of the deck as we bobbed and turned, bobbed and turned.

I looked over at my husband. He was still asleep, with his wavy black hair stuffed into the tip of the V berth. I grabbed my phone: *4:17—I shouldn't wake him.* I wanted some protection, but if it turned out to be nothing, it would spoil his first ever sleep on a rocking sailboat. I would have to investigate the intruder by myself.

I scrambled off the bed, across the hollow parquetry floor, and up the companionway stairs. I yanked back the sliding companionway cover. Disoriented by the dim dawn light, I stepped into the cockpit in my satin white pajamas. Goosebumps appeared on my forearms as bullets of rain attacked.

The wind flicked us from side to side. Just above my head, the boom swung wildly. It pulled on the mast and caused the deck to creak under the pressure. I clutched a white rope, which looked to be the mainsheet, and gave it a downward tug. The boom came to center and locked into place. The creaking stopped.

I glanced across the wet stern, which was rising and falling like a pelican in the swell. I stepped onto the port seat, from where I could see our dinghy, hiding along the port side of the sailboat. It must have moved from around the back, where we thought it would float peacefully all night. It crashed against the hull of the sailboat with every wave ... thump ... thump ... THUMP.

I scanned all 13 m of the yacht. Apart from raindrops, the deck was bare. No man to be seen. My eyes followed the waves to the brush-lined shores of Funnel Bay—no other vessels in sight.

I dragged my fingers down my damp forehead in relief. With nothing more to see, I descended into the cabin. My thigh muscles felt like gelatin. I lay down on the sofa, wedging my body between the navigation table and the galley fridge, and my heart rate finally began to slow. The blood drained from my head, and the thought I'd been avoiding for days emerged: *What the hell were we thinking, coming here?*

We chose Funnel Bay as our first night's anchorage because it was the fastest escape from land. Regulations around the movement of people to contain COVID-19 were changing from minute to minute, so our priority was to get on the water before being legally grounded. We motored full speed to Funnel Bay and arrived in the afternoon with plenty of daylight to see the reefs. We skimmed the anchoring guide of our sailing handbook and dropped anchor in the middle of the bay. It should have been a calm anchorage—it was sheltered on three sides. There were even anchor symbols on the charter company's printout, for God's sake.

The rocking of the boat was unrelenting. Over my head, the car of the self-tacking jib rushed across the deck. *It must have unfurled,* I thought. *We haven't secured that either.*

This first night aboard a sailboat,

in a gusty, rolly anchorage,

far away from everyone except my sleeping husband,

with very little control over what happened to us,

was my first formal acquaintance with the chaos that is nature.

Outside the window, the stars waved in circles. They seemed to be saying, "Yes, Jill—this is the way things have always been."

The dinghy continued to thump on the hull. The plates in the galley clinked and clinked. We bobbed and turned, turned and bobbed. And over and over like this, in no sensible pattern.

II.

I knew we needed to get out of Brisbane and as far off grid as possible in March 2020, when bedlam broke loose in Australia. The first sign cropped up in Chullora, a suburb in Sydney's outer West.

It was a crisp autumn day, and the parking lot at the Chullora Woolworths was full. Inside the supermarket, the fluorescent-lit aisles were abuzz with the movement of bright-eyed, Saturday morning shoppers.

The store was ready for Easter. Pastel yellow, blue, and pink dotted the shelves. The bakery boasted a mountain of fresh hot cross buns, steaming up the insides of their plastic packets. Near the checkout, shoppers were greeted by Lindt chocolate bunnies, each donning a pressed gold foil coat and neck ribbon with a single, miniature bell.

It was against this civilized backdrop that chaos ensued. Across the top of the toys and office supplies aisle, over the pasta and rice aisle, past the chips and candy, and all the way to fresh-cut flowers traveled the elongated squeals of a grown woman.

"Get OFF me!" she howled.[1]

The 49-year-old flailed her arms as she backed into a metal shelf.

Encroaching from both sides were a 60-year-old woman and her 23-year-old daughter.[2] If not for the age difference, they could have been identical twins. Both had jet black hair, black spandex pants, and gray jersey zip sweatshirts. Jaws clenched and eyes sharp, they approached their shopping duties that morning with sniper-like focus.

The mother and daughter homed in on the victim's T-shirt and tried to pull her away from the shelf. She was difficult to move, being bigger boned than the other two. Her stretch jeans-wrapped legs were fortresses, positioned in a wide stance with knees bent to absorb their blows. She looked as if she had been in a fight before. Her chin was tucked, and her arms swung rhythmically like medieval clubs.

The victim lunged at the two attackers, her eyes fierce. But they were possessed by the frenzy of rabid dogs. The victim's chest began to sway as she lost her footing. She screeched, "Stop! STOP!!!"

Wedged between her right ankle and the shelf was the perpetrators' target: a purple pack of Quilton 3 ply toilet paper.

Onlookers were baffled by the sight of the brawl. Even after a small crowd gathered, the mother–daughter team refused to relinquish their pursuit of the toilet paper. All the while, their trolley stood a few feet away, piled high with family-size packs of Quilton.

The store manager arrived to separate the trio. As he did, the daughter stooped to snatch the Quilton. Then she planted the pack atop the heap in their trolley.

The victim, having lost the battle, took one final stand for justice. She walked over to the perpetrators' trolley and held out her index finger before the mother attacker's nose. "I just want ... ONE ... pack!" she roared.

The mother scowled behind her glasses. Shaking her head from side to side, she answered, "No. Not one pack."

The manager observed the interaction from a distance, unsure how to handle the women. One lady at the periphery interjected, "Are you joking? Are you f—ing joking?" Other shoppers stood transfixed, as if viewing a program on the Discovery Channel.[1]

The store manager called the police, and the three were brought into the station. After questioning, the victim was released without any toilet paper compensation, while the mother and daughter received criminal charges.[3]

The story made international news.[3]

The behavior of the toilet paper-tussling women was painted as a sign of humanity's depravity. This wasn't a life-or-death situation. This was aisle 11 of Woolworths, in a country that, at the time, was ranked #7 in the world for quality of life, just behind Sweden.[4] How, the newscasters mused, could people behave so selfishly?

When 9News posted the video of the Sydney toilet paper fight on Twitter, one user commented, "We'll find a cure for coronavirus before a cure for stupid."[3] Most people would have seen the news and thrown their hands up. *That's humanity. Selfish jerks.*

But I was intrigued.

I started observing human social behavior when I was a kid because I was so very bad at it. Watching my peers through a one-way mirror, I studied every social interaction like an engineer before a new machine. I took meticulous mental notes on other people's facial expressions, the tone of their voices, and the structure of their jokes. I had volumes of research on how people behaved when they socialized, but I was always the researcher, never one of them.

If I wanted real social contact, all I had to do was step outside. Our farm in Missouri was on a long, straight road, which divided a wooded conservation park to the east and a string of corn and soy farms to the west. Roaming the pastures behind our house was a herd of Gelbvieh–Angus cattle, who had some of the sweetest eyes I'd ever seen. Each winter, flocks of geese and ducks would visit the fields next door to feed on the brome and clover that sprouted after the last cut of hay. Baby quail rummaged under a heat lamp in our garage. Once grown, they would be released into the wild. On our farm, frogs were always singing, bats patrolling, owls hooting, and blue jays swooping.

A little girl in a nature sanctuary, I dove into the lives of animals. Their world became indistinguishable from mine. Standing on the cold blades of Kentucky Bluegrass in the backyard, I would bellow "Luuuuu—keeeeee!" into the distance. Minutes later, without fail, my best friend would appear on the horizon. Lucky was a black, white, and tan Scotch Collie. He was my copilot on every farm adventure. He muddied his snow-white fur as we tromped through creeks and dragged fallen tree limbs to campsites. We spent full, tiring days exploring our terrain, returning home only with the last few rays of daylight. I would remove my mud-caked shoes and fall into a heap on the living room carpet. Lucky would collapse beside me, and I would turn to whisper my thoughts into his soft, folded ears. The rise and fall of his barrel chest was the steady rhythm on which my whole childhood hung.

Naturally, when I saw the footage of the shoppers in Sydney attacking each other over toilet paper, I didn't see humans but animals. I saw dogs fighting over a bone, curling their noses and baring their teeth. I saw cattle backed into a corner, their eyes wide with fear. But other observers didn't seem to see what I was seeing. The New South Wales police acting inspector Andrew New said of the toilet paper fight, "It's just bad behavior by people that are panicking, and there's no need for that panic."[3] To me, their behavior wasn't good or bad. They were just behaving as animals do.

Surely something under the surface was driving these females to attack each other. I grew determined to figure out what it was.

III.

For starters, I turned to another observer of animals—the grand theorist himself, Charles Darwin (1809–1882). Darwin is best known for writing *On the Origin of Species*, which directed our modern understanding of evolution. But Darwin had many other thoughts about animals prior to writing his "big book." Today we can access these thoughts via his travel diaries and over 15,000 of his letters.[5]

From the time he was a boy in rural England, Darwin was fascinated by why animals do what they do. He spent hours watching birds and collecting bird eggs, seashells, beetles, and moths.[6] Sitting in school bored Darwin—he preferred the school of nature. When he wasn't exploring the English countryside, he could be found lying under the dining room table, reading tales of the great explorers and naturalists.[6]

As a young man, Darwin fueled his dreams of distant travel with books on arctic exploration and arctic zoology. His family was wealthy and well connected to local intellectuals, giving Darwin privileged access to the latest science. He absorbed the work of French zoologist Georges Cuvier and read the work of his own grandfather, Erasmus Darwin, who proposed that animals gain certain traits through inheritance.[7]

At Christ's College, Cambridge, Darwin studied under the famous British botanist, John Stevens Henslow. Henslow was also an avid observer of nature. He liked to compare multiple specimens of the same plant species from different locations around Europe so that he could document small variations in height, leaf shape, branching pattern, or flower color.[8] These slight variations within a species were easily recognizable in Henslow's collection of over 10,000 dried plants inside the Cambridge herbarium. Given the clear evidence of natural variation within species, Henslow used to ask his students, *What defines a species?*

The natural variation in Henslow's collection at Cambridge stimulated Darwin, but he was also eager to embark on a grand voyage like the idolized naturalists of his boyhood. At age 22, Darwin was planning a voyage to Tenerife when he received a letter from Henslow with a different proposal:

> *Capt. F. wants a man (I understand) more as a companion than a mere collector & would not take any one however good a Naturalist who was not recommended to him likewise as a gentleman...The Voyage is to last 2 yrs. & if you take plenty of Books with you, any thing you please may be done—You will have ample opportunities at command—In short I suppose there never was a finer chance for a man of zeal & spirit.* (Letter from J. S. Henslow, 24 August 1831)[9]

Captain Robert FitzRoy invited Darwin to join him on a 2-year voyage charting the coast of South America. Three months later, Darwin boarded the HMS *Beagle* at Plymouth and set sail to the New World.[9]

With no clear scientific objectives for his journey through South America, Darwin began "collecting, observing, & noting anything worthy to be noted."[10] He took several specimens of octopuses, insects, and spiders,[11] recording everything that he observed in hardcover little notebooks that he always carried with him in his pocket.[12]

Following the *Beagle*'s arrival in South America, Darwin spent long stints exploring the landmass. From the ports of Rio de Janeiro, Montevideo, and Buenos Aires, he sent and received letters and shipped specimens back to Henslow.[13] In his first letter to Henslow on May 18, 1832, he confided that his head was already "full of oddly assorted ideas."[11]

A definitive moment in shaping Darwin's ideas was the occurrence of a few major natural disasters in close succession. Darwin had hoped to experience several earthquakes during his travels in South America, but according to Darwin, years passed with "only one little earthquake having happened."[14] In January 1835, the Osorno Volcano suddenly erupted in the southern Chilean Andes, and just a month later, a large earthquake shook the entire Chilean coast. After the eruption and the earthquake, Darwin and FitzRoy re-surveyed the coastline of Chile. Compared with their previous survey, the surface of the land in Concepción, the epicenter of the earthquake, had risen a small but measurable amount in altitude.[14] On a separate occasion, high up in the Uspallata pass, at 3,834 m, Darwin discovered fossilized trees that had clearly once been submerged in seawater.[14] He interpreted the fossils as further confirmation of an Earth in constant flux.

After witnessing these natural events in Chile, Darwin brooded over the dynamic nature of the Earth. If the Earth continuously remodels itself, he thought, expanding and collapsing as the molten material beneath is heated and cools, then this may explain why apparently separate phenomena, such as volcanic eruptions and earthquakes, detonate one after the other.

These great epochs of natural violence were fresh in Darwin's mind when the *Beagle* dropped anchor off the Galápagos Islands on September 15, 1835. Darwin found the Galápagos Islands to be entirely volcanic, but he also encountered some elevated deposits containing marine shells, suggesting that, like the Andes, the islands had risen from the sea long ago.[15] Darwin was surprised to see how many varieties of tortoise existed in the Galápagos, even within the same species.[16] It was just like the natural variation in plant species in Henslow's herbarium. Most of the species Darwin saw in the Galápagos were unknown on the mainland of South America and seemed to be "peculiar" to their respective environments. He admitted that it was almost as if they had been "produced on this very spot."[17] An idea was beginning to crystallize in Darwin's mind: Living things are a product of their environment. This idea was perhaps one of the most important ideas that Darwin would carry home with him from the Galápagos.

Upon his return to England, Darwin handed his collection of Galápagos plants to botanist Joseph Dalton Hooker, who found that the Galápagos plants shared many close links with the plants of mainland South America. Interestingly, though, many of the plant species existed only in the Galápagos. Of the 217 species collected, he found that 109 were confined to the archipelago and 85 were native to a single island.[18] How had entirely new species emerged on these tiny islands in the middle of the Pacific? This observation, added to years of observations in South America, stimulated fresh ideas in Darwin's mind about Henslow's question, *What defines a species?*

Eventually, Darwin's experiences and reflections would lead him to one of his most famous conclusions about the adaptation of species to their natural environments:

> *The relations just discussed ... [including] the very close relation of the distinct species which inhabit the islets of the same archipelago, and especially the striking relation of the inhabitants of each whole archipelago or island to those of the nearest mainland, are, I think, utterly inexplicable on the ordinary view of the independent creation of each species, but are explicable on the view of colonisation from the nearest and readiest source, together with the subsequent modification and better adaptation of the colonists to their new homes.* (p. 406)[19]

It was a big idea: New species are born out of previous species' adaptation to their new homes. In fact, the idea was so large, it took Darwin over 20 years from the time he returned from the Galápagos—years involving thousands of letters and discussions with prominent scientists around the world—to pull together all the puzzle pieces for *On the Origin of Species.*[20] In it, Darwin explained his holistic understanding of how living things are shaped by nature.

In nature, there is a two-way interplay between living things and the environment, which Darwin called a "struggle for existence."[19] In natural environments, living things face many challenges, such as harsh climates, limited food, limited places to nest and take shelter, predation, disease, and so on. According to Darwin, each living thing struggles against the environment, both to survive and to leave offspring.

Darwin gave a few examples:

> *Two canine animals in a time of dearth, may be truly said to struggle with each other which shall get food and live. But a plant on the edge of a desert is said to struggle for life against the drought, though more properly it should be said to be dependent on the moisture. A plant which annually produces a thousand seeds, of which on an average only one comes to maturity, may be more truly said to struggle with the plants of the same and other kinds which already clothe the ground.* (pp. 62–63)[19]

But how does our struggle for existence on this changing planet result in new species? Darwin's observations of the high degree of natural variation within a single species of tortoise or mockingbird in the Galápagos proved to be the key. Darwin theorized that because there is so much natural variation among individuals, within any population of animals or plants, there are bound to be certain individuals who are better equipped to survive dramatic changes in their environment than are others. The better adapted individuals are more likely to survive and reproduce in that particular environment, whereas less adapted individuals are more likely to die early and not reproduce. As Darwin's grandfather suggested, many of the traits that individuals possess can be passed from parent to offspring. This meant, Darwin proposed, that as generations unfold, a population of animals or plants comes to harbor the traits of only the better adapted individuals who have succeeded in reproducing. Over many generations, small differences in the traits of offspring snowball, leading to new varieties of animals and plants and eventually to new species. Darwin called this process by which new species are shaped by the environment "natural selection."[19]

Darwin was the first European naturalist to see a clear connection between Earth's dynamics and the adaptation of living forms. In subsequent years, Darwin's ideas would give rise to a revolutionary way of thinking: On a chaotic planet where the environment is always changing, animals must adapt to survive.

IV.

I remember a vivid moment in the insectary during my PhD, when a friend and fellow student turned to me, and like Mr. Miyagi, said, "Consider the grasshopper."

The grasshopper is that little green fellow who sometimes graces the balcony garden. He always comes alone. He is noble as he munches. He takes his time as he moves from one twig to the next. His every move seems judicious. It makes sense why, in Chinese culture, grasshoppers are a symbol of wisdom.

Locusts, on the other hand, are ruinous. They band together in swarms that can reach 1.5 times the area of New York City.[21] A farmer can spot a dark cloud of locusts in the sky from several kilometers away. These swarms attack crops with biblical fury. They leave no leaf unturned, blazing a path of destruction that can cost several billion U.S. dollars.[22] Because each locust can eat his weight in plants daily, in a single day, a swarm the size of Paris can eat the same amount of food as half the population of France.[23]

You wouldn't guess it by looking at them, but grasshoppers and locusts are the same bug. No, I don't mean different sub-species—they are the exact same species. Actually, there are 22 species of grasshoppers who can, at any time, turn into locusts.[21] Just like *Transformers*, the same individual can switch from a grasshopper into a locust in just a few hours.[24]

Grasshoppers normally avoid other grasshoppers as much as possible. That is, when there is enough food. But during dry spells, when vegetation dies off and only a few patches of green remain, solitary grasshoppers are forced closer together in search of food.[21] Grasshoppers sense the proximity of other grasshoppers through the hairs on their hind legs or antennae and by the sight and smell of others. These sensations of crowding sound alarm bells in the grasshopper's central nervous system. A surge of serotonin follows, launching the grasshopper's transformation into a locust.[24]

During the transformation, everything about the grasshopper's behavior and appearance changes. Increased serotonin initiates behavioral changes, such as rapid movements, varied appetite, and sociability.[21] The grasshopper's body begins to darken. No longer a cheerful green, he takes on a vicious appearance of bright yellow with black stripes. The grasshopper has transformed into a locust—a seemingly different bug.

Grasshoppers can't change their genes, so how are they able to pull this off? It turns out that within a single lifespan, grasshoppers' bodies can change which of their many genes are turned on or off, resulting in different appearances and behaviors (phenotypes) to suit new environments. This phenomenon, in which the selective expression of genes by the body produces multiple physical forms, is known as phenotypic plasticity.[25]

Beginning in the 1960s and picking up steam by the 1990s,

researchers discovered a variety of mechanisms that allow for phenotypic plasticity. In short, in species that possess this capability, only some of the body's genes are activated at once, triggering the production of certain proteins, while the other genes remain unread or "silenced." This selective expression of genes is what allows an animal to appear and act one way in one environment (mellow grasshopper) and an entirely different way in another environment (vicious locust). In other words, with phenotypic plasticity, you still have all the same keys on your piano, but you can choose which ones to play and how loudly to play them.

The flexibility to express only the genes that best suit the environment at the time is critical to a grasshopper's survival. Genetic flexibility gives grasshoppers greater behavioral breadth, allowing them to better adjust to a variety of circumstances. This is especially advantageous in a chaotic environment, which is changing all the time. It is not surprising, then, that many other animals besides grasshoppers possess such flexibility.

Responsiveness is one of the defining features of the animal kingdom. Almost all animals have a nervous system, a range of sense organs, and the capacity for some movement. These aspects of our biology allow us to respond to our environment.[26] The more chaotic the environment, the broader the range of possible conditions an animal will face and, therefore, the broader range of possible responses needed to survive.

For many animals, cooperation—defined as individuals behaving in a way that benefits other individuals—has been a mainstay of their responsiveness.[27] Cooperation has been described at all levels of biological organization.[27] Birds do it. Bees do it. Bats, bears, and baboons do it. Take the example of the yellowtail amberjack, a large fish found in the Southern Ocean. Amberjacks coordinate their actions and take specialized roles to split off mackerel from their schools and herd them toward the shore where they can attack.[28] Or cliff swallows, small colonial nesting birds that feed on insect swarms, who emit a "squeak" vocalization when they locate a swarm, calling other cliff swallows to the site.[28] Female vampire bats regurgitate blood meals to their nest mates who have failed to obtain food.[29] Adult bottlenose dolphins station themselves under other sick or injured dolphins to support their bodies at the surface so that they can breathe, leaving only when the weak animals are strong enough to swim.[30] Some cooperative tendencies in animals may be the result of genetic or epigenetic inheritance,[31] while others may be the product of individual learning or of cultural evolution, a sort of accumulation of social learning over generations.[32] Regardless of the mechanism, the result of cooperation is the same. Cooperation gives animals a greater capacity to respond to their environment. Animals who cooperate do so not out of some moral imperative but simply to survive. And because cooperation often means life or death to them, animals are some of the best teachers of cooperation available to us.

What I want to explore deeper is cooperation in one particular species: human beings. If animal behavior is already quite flexible at an individual level, it means that, if used cooperatively, it can be extremely flexible at a group level. Therefore, as a human race, we should (theoretically) be capable of adapting to an enormous variety of environmental conditions ... if we cooperate. Studies of economic games in humans show that in situations where two people could cooperate for mutual gain or sell each other out for personal gain, the most profitable strategy is to cooperate. In a strategy called "Tit for Tat," you cooperate on the first move, and you keep cooperating for as long as your partner cooperates in return. This is the most rational approach to ensure that you reap the rewards of mutual benefit without being cheated.[33] But as we saw in the Sydney toilet paper fight, humans don't always do what's rational. In times of chaos, when people most need to cooperate, we often don't. Why?

In this book, we'll untangle how a trait like cooperation might have evolved in humans. Darwin showed that humans are interdependent not only with each other but with nature. And because nature is in constant flux, nature often rattles us, like the weather in Funnel Bay rattled me. Instead of causing our species to give up and die out, our "struggle for existence" amidst nature's dynamics has shaped us in another direction entirely—one in which we've developed flexibility and resilience by working together.

To understand the role of environmental chaos in the evolution of cooperation, we'll continue down the path of inquiry that Darwin started for us. Accordingly, each chapter of this book is presented as a little notebook, much like the ones Darwin kept. Throughout the course of our notebooks, we'll consider how Earth's dynamics could have shaped the cooperative behaviors of social animals and early humans, beginning at our divergence from the other great apes. Cooperation is a behemoth of a topic, and writing the definitive book on human cooperation is virtually impossible. But by treading in Darwin's footsteps, together we can begin to sift through this "strange assemblage of ideas."[15]

The future of humanity is in our hands. What we learn from other animals and our early ancestors can give us a better understanding of cooperation and what it means for the survival of our species. I believe that, as humans, we can use our innate cooperative behavior to navigate through the chaos before us and respond to humanity's greatest challenges. But to do so, we'll need to make careful observations and take good notes, all the while asking:

What is the glue that binds us as a species,
and how do we use it for good?

CHAPTER 1 REFERENCES

Free Resources

Online compendium of studies on the evolution of cooperation, put together by Robert Axelrod and Lisa D'Ambrosio: http://www-personal.umich.edu/~axe/research/Evol_of_Coop_Bibliography.htm

More information about Charles Darwin's life, including a searchable database of excerpts from the 15,000 letters that he wrote, can be found at Cambridge University's Darwin Correspondence Project: https://www.darwinproject.ac.uk/

Further Reading

Clutton-Brock, T. (2009). Cooperation between non-kin in animal societies. Nature, 462(7269), 51–57. https://www.nature.com/articles/nature08366

Trivers, R. L. (1971). The evolution of reciprocal altruism. The Quarterly Review of Biology, 46(1), 35–37. https://pdodds.w3.uvm.edu/files/papers/others/1971/trivers1971a.pdf

References

1. Falanga. (2020, March 7). *Fight erupts in Woolworths over the last pack of toilet paper* [Video]. YouTube. https://www.youtube.com/watch?v=eAO5XfVionE

2. "It isn't Mad Max": Women charged after fight over toilet paper in Sydney. (2020, March 8). *The Guardian*. https://www.theguardian.com/australia-news/2020/mar/07/it-isnt-mad-max-police-warning-after-shoppers-brawl-over-toilet-paper-in-sydney

3. Brown, L. (2020, March 8). Women fight over toilet paper during coronavirus buying in Australia. *New York Post*. https://nypost.com/2020/03/08/women-fight-over-toilet-paper-during-coronavirus-panic-buying-in-australia

4. US News. (2019). *Best countries overall rankings 2019*. https://www.usnews.com/media/best-countries/overall-rankings-2019.pdf

5. University of Cambridge. (2022). *Darwin's life in letters*. Darwin Correspondence Project. https://www.darwinproject.ac.uk/letters/darwins-life-letters

6. American Museum of Natural History. (2005). *Young naturalist*. https://www.amnh.org/exhibitions/darwin/young-naturalist

7. University of Cambridge. (2022). *Darwin's student booklist*. Darwin Correspondence Project. https://www.darwinproject.ac.uk/people/about-darwin/what-darwin-read/darwin-s-student-booklist

8. Kohn, D., Murrell, G., Parker, J., & Whitehorn, M. (2005). What Henslow taught Darwin: How a herbarium helped to lay the foundations of evolutionary thinking. *Nature, 436*(4), 643–645.

9. University of Cambridge. (2022). *Voyage of HMS Beagle.* Darwin Correspondence Project. https://www.darwinproject.ac.uk/commentary/voyage-hms-beagle

10. Cambridge University Library. (2009). *"Not a finished naturalist."* A Voyage Round the World. https://wwwe.lib.cam.ac.uk/CUL/exhibitions/Darwin/naturalist.html

11. Darwin, C. (1832). *To J.S. Henslow, 18 May–16 June 1832.* Darwin Correspondence Project. https://www.darwinproject.ac.uk/letter/DCP-LETT-171.xml

12. Kennedy, M. (2009, November 24). Appeal launched for Charles Darwin's Galapagos notebook. *The Guardian.* https://www.theguardian.com/science/2009/nov/24/appeal-charles-darwin-galapagos-notebook

13. Cambridge University Library. (2009). *"A cruize in these latitudes."* A Voyage Round the World. https://wwwe.lib.cam.ac.uk/CUL/exhibitions/Darwin/cruize.html

14. University of Cambridge. (2022). *Darwin's earthquakes.* Darwin Correspondence Project. https://www.darwinproject.ac.uk/commentary/geology/darwin-s-earthquakes

15. Chancellor, G., & Keynes, R. (2006). *Darwin's field notes on the Galapagos: "A little world within itself."* Darwin Online. http://darwin-online.org.uk/EditorialIntroductions/Chancellor_Keynes_Galapagos.html

16. Chancellor, G., & van Wyhe, J. (2008). *"Ladies, like mermaids": An introduction to the Galapagos Notebook.* Darwin Online. http://darwin-online.org.uk/EditorialIntroductions/Chancellor_fieldNotebooks1.17.html

17. Darwin, F. ed. (1909). *The foundations of the origin of species: Two essays written in 1842 and 1844.* Cambridge University Press. https://darwin-online.org.uk/content/frameset?itemID=F1556&viewtype=text&pageseq=1

18. Darwin, C. (1845). *Journal of researches into the natural history and geology of the countries visited during the voyage of H.M.S. Beagle round the world, under the Command of Capt. Fitz Roy, R.N.* (2nd ed.). London: John Murray. https://darwin-online.org.uk/content/frameset?viewtype=text&itemID=F14&pageseq=1

19. Darwin, C. (1859). *On the origin of species by means of natural selection, or the preservation of favoured races in the struggle for life.* London: John Murray. https://darwin-online.org.uk/content/frameset?itemID=F373&viewtype=text&pageseq=1

20. University of Cambridge. (2022). *The writing of "Origin."* Darwin Correspondence Project. https://www.darwinproject.ac.uk/writing-origin

21. Maggiacomo, T., & Berne, K. L. (n.d.). How to set off a plague of locusts. *National Geographic.* Retrieved April 26, 2023, from https://www.nationalgeographic.com/magazine/graphics/how-to-set-off-a-plague-of-locusts-interactive

22. The Food and Agriculture Organization. (n.d.). *Countries take responsibility for regional desert locust control.* Retrieved April 26, 2023, from https://www.fao.org/3/i2940e/i2940e17.pdf

23. Locusts: Photo ark. (n.d.). *National Geographic.* Retrieved April 26, 2023, from https://www.nationalgeographic.com/animals/invertebrates/facts/locusts

24. Rogers, S. M., Matheson, T., Sasaki, K., Kendrick, K., Simpson, S. J., & Burrows, M. (2004). Substantial changes in central nervous system neurotransmitters and neuromodulators accompany phase change in the locust. *Journal of Experimental Biology, 207*(20), 3603–3617. https://doi.org/10.1242/JEB.01183

25. Simpson, S. J., Sword, G. A., & Lo, N. (2011). Polyphenism in insects. *Current Biology, 21*, 738–749. https://doi.org/10.1016/j.cub.2011.06.006

26. Broom, D. M., & Johnson, K. G. (2019). *Stress and animal welfare: Key issues in the biology of humans and other animals* (2nd ed.). Springer. https://doi.org/10.1007/978-3-030-32153-6

27. van Schaik, C. P., & Kappeler, P. M. (2006). Cooperation in primates and humans: Closing the gap. In *Cooperation in primates and humans: Mechanisms and evolution* (pp. 3–21). Springer. https://link.springer.com/chapter/10.1007/3-540-28277-7_1

28. Schmitt, R. J., & Strand, S. W. (1982). Cooperative foraging by yellowtail, *Seriola lalandei* (Carangidae), on two species of fish prey. *Copeia, 3*, 714–717. https://www.jstor.org/stable/1444679

29. Wilkinson, G. S. (1984). Reciprocal food sharing in the vampire bat. *Nature, 308*(5955), 181–184. https://doi.org/10.1038/308181a0

30. Siebenaler, J. B., & Caldwell, D. K. (1956). Cooperation among adult dolphins. *Journal of Mammalogy, 37*(1), 126–128. https://doi.org/10.2307/1375558

31. Trerotola, M., Relli, V., Simeone, P., & Alberti, S. (2015). Epigenetic inheritance and the missing heritability. *Human Genomics, 9*(1), 17. https://doi.org/10.1186/s40246-015-0041-3

32. Whiten, A. (2019). Cultural evolution in animals. *Annual Review of Ecology, Evolution, and Systematics, 50*(1), 27–48. https://doi.org/10.1146/annurev-ecolsys-110218-025040

33. Axelrod, R., & Hamilton, W. D. (1981). The evolution of cooperation. *Science, 212*(4489), 1390–1396. https://doi.org/10.1126/SCIENCE.7466396

CHAPTER 2

Blowin' in the Wind: Uncertainty

I.

In the days preceding our decision to escape to the Whitsundays, the air in our neighborhood streets of Brisbane grew stiffer as the headlines became graver. My husband and I, always quick to criticize the media for igniting paranoia, believed that like all so-called "crises," this coronavirus thing would soon blow over.

While others grew wary of being in public, we upheld our midweek custom of going down the street for a cold beer on tap. Our basset hound, Leo, was excited by the outing. He kept a taut leash and a pep in his trot all the way down the sidewalk. Daylight was fading as we approached the corner. I glanced across the empty tables and chairs of the Vietnamese bar and bistro. In an instant, fairy lights on half a dozen trees along the main street came alive.

We walked up to the bar and settled into the tropical-foliage-print cushions on the patio chairs. The bar was dark, aside from the overly saturated images that flashed on an LED screen mounted above it. It was the third Wednesday in March—we were meant to go scuba diving in Mexico in just a few days. We had finished our Advanced Open Water Diver course 30 months earlier, and we'd plunged into Mexico's turquoise waters enough times to seed an addiction.

Our eyes gravitated to the TV screen, where rolling headlines and moving mouths provided weak signals as to whether our Qantas flight to Los Angeles would go ahead.

Tension had been swelling in distant pockets of the globe for nearly 3 months. In mid-December 2019, a few hospitals in central China had begun to notice multiplying presentations of patients with pneumonia-like symptoms. Then, on December 30th, 33-year-old ophthalmologist Li Wenliang transmitted a private message over WeChat to a group of his fellow doctors warning them about a possible viral outbreak in his city, Wuhan, in the Hubei province. He urged them to do what they could to protect themselves from getting the infection.[1, 2, 3] Eleven days later, he developed a cough.[4]

On January 11, 2020, Chinese state media released an official report of the first death from a novel coronavirus, called SARS-CoV-2.[5, 6] The deadly flu-like illness caused by the virus would become known as COVID-19.[6] On the 13th of January, a 73-year-old Thai woman who had visited Wuhan in December came down with a severe cough.[7] Just four days later, a case emerged in France. Soon after that, a cluster of cases appeared in Haute-Savoie, France, and another cluster in Bavaria, Germany.[8] By the 20th of January, people had carried the virus to the United States, Japan, and South Korea.[9]

On the 7th of February, the young ophthalmologist from Wuhan, Li Wenliang, died.[1]

Global hysteria took hold as the death toll rose. Italy confirmed their first case of local transmission in the northern alpine region on the 21st of February. Within 19 days, Italy had 12,462 coronavirus patients, more than 1,000 people in intensive care, and 827 dead.[10] On March 9th, the Italian Prime Minister, Giuseppe Conte, imposed a nation-wide lockdown, asking people to stay home to stay safe. He closed all public gatherings, schools, shops, and restaurants after dusk. In his address to Italy's 60 million citizens, he quoted Winston Churchill, calling the situation Italy's "darkest hour." He went on to admit, "Non c'è più tempo," which translates as, "We have run out of time."[11, 12]

No leader in the world knew what to do to stop the virus. But with over 14 million people flying on any given day,[13] shutting down international air travel was an obvious start. Italy enacted a full ban on flights from China after the first Chinese couple tested positive in Rome. But Italy's borders with other countries remained open, and Australian politicians were quick to point out that Italy's cases were rising.

My husband pinched to zoom on the flight radar app open on his phone, which lay on the bar's tempered glass table. Several flights were still going over the Pacific.

On the TV above the bar, the regular broadcast was interrupted by the appearance of Australia's Prime Minister, Scott Morrison. His round head, dusted with wispy gray hair and mounted on a starched white collar, looked like a marble bust. My husband and I went into a hush. With the furrowed brow of a school principal, the prime minister stared at us over the wire glasses that rested low on his nose.

"Stop hoarding," he said.[14]

He said it in the same way an exasperated father warns his 10-year-old son, *Stop hitting your sister.*

He went on. "I can't be more blunt about it. Stop it."[14]

I nearly choked on my beer. *Was this guy serious?*

"It's not sensible. It's not helpful," he said. "And I've got to say, it's been one of the most disappointing things I've seen in Australian behavior in response to this crisis."[14]

Personally, I would have thought that his government's frazzled response—reminiscent of a middle-aged woman grasping for dollar bills inside a money blowing machine—was more disappointing. Thankfully no one asked me.

"That is not who we are as a people," he said. "It is not necessary. It is not something that people should be doing. ... There's no reason for people to be hoarding supplies."[14]

Denying basic animal emotions, I mused. *Great job, ScoMo*. My ears tuned out for the rest of his speech. He obviously had no idea of who we are "as a people."

The lanky, young bartender piled dishes into a plastic tray and lugged them back to the dishwasher. He proceeded to rinse down the bar. My husband and I sipped our beers, the topography of our faces harshened by the glow of the screen. Done with his tasks for the night, the bartender came around to the front of the bar to get a closer look at Leo, who was sprawled on the floor.

He squatted and massaged the dog's jelly-like, 28 kilogram figure. We asked how business had been lately. I don't love Vietnamese food, but I've always liked the idea of having someplace open at the end of the street, if anything, just as a sign of life.

"People have stopped coming," said the young man, his voice breaking. "We still do a lot of takeaway, but not anything like before." I nodded in an attempt at empathy, although it had been over a decade since I had worked in a restaurant. He caved his chest and shoulders over Leo further, moving his hands to the basset's velvety ears. Then he looked up at us, his eyes watery and resigned. "A lot of people have stopped going to Asian restaurants," he said, "because people know the virus started in China."

The final "a" of "China" hung in the air.

I froze.

"Wow," I said. He nodded slowly, turning his gaze across the empty street, where the warm fairy lights on the trees now seemed out of place.

"People are scared," he said. "I don't think people need to be scared, but they are."

He was right.

Having spent most of my life around animals, I'm highly attuned to animal behavior. What leaders like ScoMo seemed to be forgetting—*Stop hoarding ... stop it!*—is that humans, like other animals, have evolved to make life-saving decisions in the face of threats. But we depend on some base level of certainty to make good decisions.[15] If there's one thing that the Sydney toilet paper fight demonstrates, it's not how selfish humans are. It's how easily uncertainty overcomes us, precisely when we need to cooperate the most.

II.

After ScoMo's speech on Wednesday, I knew we had to get out of Brisbane—fast. Having come of age in Middle America in the 1990s, I had witnessed how uncertainty can metamorphose into full-blown societal paralysis. I didn't need to stick around to know what was about to happen to a society in the grip of fear.

The early '90s in the United States held a lot of promise for a kid like me. The president was cool. The suburbs were safe. The music was shameless. The public saw the birth of the World Wide Web. Within the span of a few years, the American middle class gained access to space-age technologies, from the mobile flip phone to dial-up internet. Our future seemed certain; it seemed neon bright.

But in 1999 our certainty began to topple. On a Tuesday in April during lunch break, two twelfth-grade students showed up at their school with duffle bags of improvised explosives. The school was 9 hours west of us in Columbine, Colorado. After the bombs planted in the cafeteria failed to detonate at 11:17 a.m., the two twelfth-graders uncovered semi-automatic pistols from their black trench coats and began shooting. They proceeded through the school to murder 10 other students.

What I heard about the Columbine library was especially disturbing to me—as an introvert, I had always considered the school library to be my haven. One of the Columbine shooters walked over to a library table, under which a 17-year-old girl was hiding. He slapped the surface of the table twice, and he knelt down to see her. He said, "Peek-a-boo." Then he shot her in the head.[16]

Across the United States, our illusion of safety vanished. The old wooden doors throughout my elementary school were replaced with heavy, bullet-proof ones. We started holding active shooter drills, in which the teacher told us to stay silent and hide out of view while she locked the classroom door from the inside. Every visitor to our school was treated with suspicion. We began to treat each other with suspicion too. I added that weird, disheveled kid at the back of class, who drew impaled bunnies in black and red ink, to my personal watch list.

Today we understand more about what uncertainty does to humans. Prior to the Vietnam War, soldiers who suffered mental distress following traumatic experiences on the battlefield were given the diagnosis of "soldier's heart," "shell shock," or "war neurosis." It was only after a quarter of all soldiers sent to Vietnam came home with psychological distress that the American Psychiatric Association recognized a unique kind of anxiety that was present in some soldiers after war.[17] Post-traumatic stress disorder (PTSD) was the new diagnosis given to people who had been through a traumatic experience, like military combat, and later suffered chronic recurrent thoughts of the trauma, went to great lengths to avoid anything that resembled the trauma, and remained in a state of hypervigilance long after the trauma, always on the lookout for threats.[17]

At first, psychiatrists thought that PTSD was a type of Generalized Anxiety Disorder (GAD), but studies over the last four decades have found unique biological mechanisms underlying PTSD. Both GAD and PTSD involve the amygdala, a brain structure that participates in our learned responses to fear. The difference with PTSD is the element of uncertainty. For instance, when researchers used a predictable or an unpredictable schedule to play aversive sounds, like a smoke alarm or a woman screaming, to people with PTSD or GAD, the participants with PTSD startled much more than those with GAD when the timing of the sounds was unpredictable.[18]

The behavioral differences between GAD and PTSD can be traced to differences in brain activity. Recent brain studies have found that GAD involves hyperactivity of the medial division of the central nucleus of the amygdala, which coordinates rapid fear responses to imminent stressors that are relatively certain, whereas in PTSD, the hyperactivity occurs in the bed nucleus of the stria terminalis, a region activated under conditions of a sustained, unpredictable threat.[15, 19, 20, 21] Under normal circumstances, the bed nucleus of the stria terminalis helps us to distinguish between safe and unsafe contexts.[22] But in war, where threats can emerge at any time without warning, this region locks the brain into vigilance overdrive, so the soldier is constantly ready for any kind of attack.[20]

Findings in soldiers with PTSD align with findings in animals in the presence of predators. In other mammals, such as rats, the central nucleus of the amygdala that is overactive in GAD appears to be associated with the detection of predators at close range and activation of the "fight-or-flight" response. Meanwhile, the bed nucleus of the stria terminalis, involved in PTSD, is associated with the detection of predators at long range, when the nature of the threat is uncertain.[15, 20, 23]

This parallel between humans and other animals may explain why people with PTSD react to an imminent threat just like anyone else would, with a classic fight-or-flight response, but they overreact to distant or uncertain threats.[24, 25] The innate animal response to uncertain threats is to freeze or hide. Freezing is a better strategy for protection when a predator is far away, whereas the fight-or-flight response is better when the predator is near. With many "far away" threats in their awareness, war veterans who develop PTSD continue to have a freeze response to uncertainty that exceeds what is needed in peacetime.

Prolonged uncertainty can also alter how we assess risks.[15] War veterans with PTSD tend to see things as more negative than they are and overestimate the probability that something bad will happen, a phenomenon referred to as cognitive bias.[26] Cognitive bias is also a test used by animal welfare scientists to determine whether animals are in a chronic, poor mental state.[27] For instance, studies of judgment bias tasks in rats (*Rattus norvegicus*), Atlantic bottlenose dolphins (*Tursiops truncatus*), western lowland gorillas (*Gorilla gorilla gorilla*), and an American black bear (*Ursus americanus*) have found that animals experiencing poorer welfare will judge ambiguous stimuli more pessimistically.[28, 29] Although uncertainty is something that all animals must cope with from time to time, studies of war veterans and animals experiencing poor welfare have taught us that too much uncertainty for too long can chronically warp the way in which we see the world.

Before school shootings became common, Americans had enough threats to worry about, like kidnappings and drive-by shootings. But these threats were somewhat predictable. Kidnappings are more likely when kids are left unsupervised, and drive-by shootings tend to occur in neighborhoods where other crimes abound. Add to these threats the unpredictability of school shootings, however, of never knowing when or where an active shooter might let loose on our innocent children sitting in the classroom, and Columbine was enough to plunge Americans into a snap-freeze of anxiety for the arrival of the new millennium.

I thought nothing could be more unsettling for Americans than Columbine, but I was wrong. At 8:46 a.m. on September 11th, 2001, five hijackers crashed American Airlines Flight 11 into the World Trade Center's North Tower in the heart of New York City. Seventeen minutes later, another five hijackers crashed United Airlines Flight 175 into the South Tower. At 9:37, American Airlines Flight 77 crashed into the Pentagon, and at 10:03, United Airlines Flight 93 crashed near Shanksville, Pennsylvania, after the passengers attempted to overtake the hijackers.[30]

Following the September 11th attacks, the United States went into turmoil for hours, days, and months. New York City firefighters, police officers, construction workers, search-and-rescue dogs, and volunteers worked around the clock in attempts to rescue people trapped under the debris of the World Trade Center. Careful not to disturb the precarious rubble, at first some of them dug with their bare hands. In what has been referred to as a "disaster marathon,"[31] rolling news coverage of the attacks occupied the TV screen for 5 days straight, as the planes crashing into the World Trade Center's buildings replayed on loop.[32]

Reverberations of 9/11 on the American psyche were long lasting. Fires burned at Ground Zero for 100 days after the attack, creating a black smoke cloud that could be seen from New Jersey. Cleanup and recovery efforts of 1.6 million tonnes of wreckage lasted more than a year.[33] With 2,996 fatalities, it was the largest loss of life by a foreign attack on American soil. It was the deadliest terrorist attack in human history.[34] And for many Americans, it signaled an end to safety within our own borders. The unpredictability of another terrorist attack would haunt us for decades. I believe that, as a society, we are still on guard now. We are still looking for threats even where there are none. We are chronically anxious—lost in a sea of uncertainty.

III.

One of the best ways in which animals mitigate uncertainty is by living in groups.

Many animals spend the majority of their lives in social groups: bees, elephants, lions, buffalo, meerkats, chimpanzees, gorillas, penguins, hyenas, mice, bats, swallows, and dolphins, just to name a few. The distinction between group and solitary life is fuzzy, because even solitary animals engage in social interactions at some point in their lives (e.g., mating, rearing young, or exploiting a large food source like the annual migration of prey). Animals who live close together for a long time, however, are said to live in groups.[35]

Many evolutionary biologists believe that solitary life is the ancestral state of life—that our oldest animal ancestors would have lived alone. Group living is believed to have evolved only later, having emerged independently across various animals when ecological conditions made it so that the benefits of group living outweighed the costs.[36]

Group living is not an easy way of life. It comes with a host of problems for animals. The most pressing problem is that as groups grow larger, they exhaust patches of food more quickly, so it's harder for each individual to find food. To compensate for this, individuals either have to spend a larger portion of their day looking for food or they have to travel farther to find new food patches.[37] Researchers have found that larger groups of primates travel farther due to their faster food depletion. Perhaps because of this difficulty in securing enough food for everyone, female primates often bear fewer offspring on average when living in larger groups.[38] Apart from the challenge of finding food, animals living in groups are also more likely to have any food that they have secured stolen from them by a fellow group member. Contests over food between group members can quickly escalate to aggression, injuries, and death. In addition, animals living in tight-knit groups are more prone to getting parasites and diseases from each other, which can slow their growth, reduce their reproduction, or cut their lives short. Inbreeding is also more common in group-living animals, as is infanticide, in which one's offspring are killed by other group members.[36]

Charles Darwin insisted that for animals to live in groups, there must be some benefit for their survival or reproduction. In *On the Origin of Species*, Darwin suggested that for a trait to persist in a population through natural selection, different individuals in the population must have different traits (phenotypic variation). What's more, those different traits must confer different advantages for survival and reproduction in different environments (differential fitness), and the traits must be able to be passed from generation to generation (heritability).[39] In his 5th edition of *On the Origin of Species*, Darwin described natural selection using English polymath Herbert Spencer's term "survival of the fittest,"[40] with "fittest" referring to those individuals who are better suited to their immediate, local environment.[41]

Given the major disadvantages to survival and reproduction, group living must offer animals an equally compelling set of advantages. The most obvious benefits of group living include better access to resources through group hunting and sharing information; better safety from predators through increased vigilance and joining forces to drive predators away; and better access to mating opportunities and shared responsibility for rearing young.[36] However, there are many other less obvious benefits of living and cooperating in groups that we are only starting to understand, such as the psychological benefits of social bonding.

If we focus only on the costs and benefits to the individual, however, we are missing a large part of the picture. Most fitness models to date have looked at why individuals benefit from living in groups. Confusingly, though, we see in nature that many primate group sizes are far larger than what models indicate as optimal for individuals.[37] So how would Darwin explain the existence of such large groups?

Darwin himself struggled to understand how certain characteristics of animal groups could have emerged according to his theory of evolution. For example, when members of a group do things for each other, sometimes the behaviors are advantageous to the group but not necessarily to the individual. How could a trait like cooperation in groups persist, Darwin wondered, if it doesn't help the individual?

In honey bee colonies, only the queen bee can produce offspring. The other female bees are sterile. These other females serve the queen bee to build the hive and support the colony, but they cannot pass on any of their own traits. How could this self-sacrificing trait have emerged in the first place, Darwin wondered, if the worker bees are sterile? He admitted that the case of worker bees who sacrifice themselves to protect their hives was "one special difficulty, which at first appeared to me to be insuperable, and actually fatal to the whole theory"(p. 236).[42]

After much contemplation, Darwin finally realized that the problem of worker bee cooperation "disappears when it is remembered that selection may be applied to the family, as well as the individual..."(p. 204).[42] In other words, if we view the entire bee colony as one big family, then by supporting the queen bee's reproduction, the worker bees are supporting their kin. Because the worker bees are genetically related to the queen bee, their support of the queen bee's reproduction ensures that some of their own genes are passed on to the next generation. This premise eventually became known as "kin selection."[43] Evolutionary biologist William D. Hamilton put mathematics to kin selection, interpreting that cooperative behavior can contribute to an individual's "inclusive fitness." Even when cooperation doesn't directly benefit the individual, if it's not too costly to do and it improves the fitness of one's close relatives, then it will persist.[44] Kin selection may explain why families of animals cooperate, even when it puts an individual at risk, and why humans will help our relatives, even when it hurts.

What kin selection doesn't explain is why *unrelated* individuals, like Darwin and his mentor, Henslow, sometimes cooperate. In fact, the inexplicability of human cooperation was the major sticking point for Darwin with his whole theory of evolution. Darwin suggested that perhaps, "as the reasoning powers and foresight of the members became improved, each man would soon learn from experience that if he aided his fellow-men, he would commonly receive aid in return"(p. 163).[45] In this explanation, known as "reciprocal altruism," acts of altruism can be seen as a form of enlightened self-interest, where it pays to be nice to others because at some point in the future, they will be nice to you in return.[46] Despite the possibility of reciprocal altruism, the problem of human cooperation continued to trouble Darwin in his later life. In the *Descent of Man*, Darwin proposed yet another explanation:

> *It must not be forgotten that although a high standard of morality gives but a slight or no advantage to each individual man and his children over the other men of the same tribe . . . an increase in the number of well-endowed men and an advancement in the standard of morality will certainly give an immense advantage to one tribe over another.* (p. 166)[45]

Darwin reasoned that although selfish individuals might out-compete altruistic or self-sacrificing individuals within the same group, altruistic groups will likely out-compete selfish groups.[47] Over a century later, sociobiologists Edward O. Wilson and David Sloan Wilson turned Darwin's logic into a handy rule of thumb: "Selfishness beats altruism within single groups. Altruistic groups beat selfish groups."[47]

Darwin remained convinced that on almost all matters of evolution, natural selection occurs at the individual level; that is, individuals who are better suited to their local environment are more likely to survive and reproduce than poorly suited individuals.[48] But his provocative suggestion that natural selection may sometimes operate at higher levels of organization, selecting for fitter families or fitter groups, inspired later evolutionary biologists to elaborate on this theory. Eventually it became known as multilevel selection theory.[47]

Multilevel selection theorists conceptualize the total evolutionary change in a population as a giant arrow made up of component arrows, which are sometimes pointing in opposite directions. What is advantageous to the individual or the individual's genetic relatives (within-group selection) may be different to what is advantageous to the group or the broader population (between-group selection); but it is the sum of these advantages that matters for natural selection.[47] Of course, natural selection at higher levels does not replace or eliminate selection at the individual level. But by focusing on multiple levels of selection rather than on the individual alone, we can gain a better understanding of sociality in humans and many other animals.[47]

In light of Darwin's realizations about group living, it may be useful to consider the advantages that group living brings to the group, rather than just to the individual. Living in groups provides animals with a kind of buffer against an ever-changing, uncertain environment. Group living is a way of overcoming what Edward O. Wilson called "phylogenetic inertia," or the evolutionary constraints that make it impossible to change one's own biological traits quickly enough in order to adapt to changes in the local environment.[49, 50] According to this theory, even if our bodies are incapable of changing quickly enough, cooperation can give us (as a group) the behavioral flexibility to adapt and survive in many environments. Moreover, we all know that some burdens in life are just too heavy to bear alone. A cooperative group may be able to tolerate higher levels of external stress, or allostatic load, than any one individual could tolerate.[37] For these reasons, cooperation can ensure the flexibility and endurance of a group. When faced with an uncertain environment, groups who can cooperate may have a survival advantage over groups who cannot cooperate. When cooperative groups survive, they pass on their prosocial genes[51] and cooperative cultures,[52] thereby seeding cooperative behaviors in the generations to come.

IV.

For humans, group living is by far the most common way of life. It's no wonder that little acts of cooperation, like picking up an item that someone else has dropped, often feel like second nature to us. Some evidence suggests that cooperation is instinctual for humans. We do it almost from birth, without having to be taught how to do it. Infants under the age of one year have been found to prefer puppets that are helpful to others over puppets showing antisocial behavior,[53] indicating that our predisposition to cooperate may be hardwired. Studies of human decision-making during economic games suggest that our tendency to cooperate is a product of our intuition, rather than the result of rational reflection.[54] It could be that cooperation is deeply ingrained in us and is, as Darwin suspected, a product of our evolution.

The evolution of human cooperation likely stretches back at least 6 million years to our last common ancestor with other great apes.[55] Today the great apes are a family of primates that includes three genera of group-living animals—humans, gorillas, and chimpanzees—and the semi-solitary genus of orangutans. But when they first emerged during the Miocene epoch, which was a warmer than usual time on Earth 23–5.3 mya, the great apes were much more abundant, diverse, and widespread than they are today.[56] This Miocene explosion in apes is what paleoanthropologist David Begun has called "the real planet of the apes."[57]

The Miocene was an uncertain time to be a primate. Carbon and benthic oxygen isotopes show there were extreme environmental fluctuations in Africa from the middle Miocene onwards.[58, 59] Rainfall varied wildly during the Miocene, as seen in the abundance of calcareous nodules deposited in the soil during long dry seasons and the waterlogged, oxygen-depleted soil of the wet seasons.[59] Between 17 and 13.5 mya, the Earth's mantle shifted upwards, creating the East African Plateau.[60] The resulting Great Rift Valley formed a chain of volcanoes that are still active today.[60] From around 16 mya, these volcanoes erupted, remodeling the landscape of East Africa.[61]

Our great ape ancestors were forced to adapt to a mosaic of new habitats during the Miocene. The uplift of the East African Plateau represented a new land barrier to moisture-bearing winds, which created a new patchwork of humid and arid environments.[62] From the middle to late Miocene, many of the once-abundant tropical forests transitioned to subtropical woodlands. Unlike the lush forests we see in central Africa today, these dry woodland habitats had relatively open canopies and were home to fewer species.[59] In the late Miocene, it might have been advantageous to be an ape with greater behavioral flexibility, one capable of moving between trees and over ground and exploiting a variety of habitats.

It was during these intense habitat changes of the late Miocene that our last common ancestor with other great apes lived. Our last common ancestor was, more than likely, a cooperative ape. Although scientists once believed that the evolution of sophisticated cognition, including "theory of mind," must have happened before the advent of cooperation, many scientists now believe the reverse is true.[63] Cooperation in groups likely happened first, before humans diverged from other great apes some 6 mya, and this was followed by the evolution of basic cognitive abilities, such as the ability to point things out to others and to laugh along with others in a group.[63, 64, 65] It could be that in the uncertain times of the Miocene, those apes who could cooperate with others in groups were more resilient to environmental changes and were therefore better able to survive. The great apes who remain today, including humans, are the direct descendants of those cooperative apes.

Most great apes, including humans, have evolved over millennia to cooperate in the face of environmental uncertainty. In today's world, however, we are starting to forget our past. As social animals, we have an incredible ability to cooperate, endowed to us by millions of years of environmental uncertainty. But we're no longer using our cooperative abilities to their fullest potential. Instead, we sit frozen before the unknown.

I was recently in a room with about a hundred Australian farmers and agriculture specialists, all discussing the impacts of climate change on farming. Going forward, they wondered, how will farmers deal with the extremes in temperature and rainfall that are transforming the farming industry into a Russian roulette? You could see the helplessness on their faces.

Our globally connected markets and societies place us at even greater risk of food insecurity. Major disruptions like wars and pandemics affect all four pillars of food security: (1) Is the supply of food adequate? (2) Can people obtain the food they need? (3) Do people have a sufficient intake of nutrients? (4) Can people access food at all times?[66] These are concerns that our great ape ancestors in the Miocene would have had to deal with too. Their way of adapting was to cooperate.

Food security is only one facet of the uncertainty facing humanity. There are extreme weather events, natural resource depletion, dwindling biodiversity, financial volatility, demagoguery, racism, extremism, and global pandemics. These are all complex problems with no clear-cut solutions. Overwhelmed, we are prone to freeze before them. I believe that, as a species, we are experiencing increasing anxiety, because we didn't evolve to tackle such problems as individuals. We didn't progress this far by working alone.

I worry that we've become overwhelmed with uncertainty. Ever since that dramatic crash in optimism at the start of the millennium, anxiety in human populations around the world has skyrocketed. We are beginning to realize that the fragile interdependence of life forms on our planet makes our world practically unpredictable. Even after COVID-19 no longer threatens millions of lives, there will be other environmental threats, perhaps even greater ones. At times, many of us (myself included) have chosen to bury our heads in the sand. It's only natural, when faced with uncertainty, for us to freeze. But the risk for society if we continue to strengthen these neural circuits is that, like a soldier with PTSD, we'll interpret our world as more and more threatening, and our anxiety will leave us too frozen to respond.[67]

At a time when humans are most uncertain, cooperation just might be the way forward. But it isn't going to be comfortable. Cooperation involves risk. It would certainly be more comfortable for us to continue waiting for a perfect solution to appear, trusting blindly in our elected leaders, all while blaming one another and fueling conflict. What would be uncomfortable but prudent is if we humans started to "stumble through" the uncertainty together. Like soldiers supporting each other through a dense fog, we could take on the threats in front of us together. As a species, collective stumbling may be as good as we're going to get.

I think, at heart, we all know there is no canned solution to our problems. If the solution to all of humanity's problems were a perfect circle, we would be lucky to even approximate it with a poorly drawn polygon. The more people who work on a problem, however, the more sides we can add to the polygon, and the closer it will approximate a circle. I suppose what we need to decide is whether we want to hold out for that perfect circle to fall from the sky, or whether we start working on the polygon now.

No one is saying this will be easy. But when you're in a precarious predicament, as humanity is now, it would be wise for us to seriously consider cooperation—a behavior that has ensured the survival of our forebears for the last 6 million years. Navigating through the chaos in front of us is always going to be difficult. But thankfully, as a species, we have a long and rich history of working together on difficult things.

I'm thrilled to have you join me on this adventure of uncertainty, discovery, and action. After all, the best ideas and most successful plans will not be mine or yours or hers or his, but *ours*.

CHAPTER 2 REFERENCES

Free Resources

van Wyhe, J. (2002). *The complete work of Charles Darwin online*. http://darwin-online.org.uk/

Further Reading

Sussman, R., & Cloninger, C. (Eds.). (2011). *Origins of altruism and cooperation*. Springer. https://doi.org/10.1007/978-1-4419-9520-9

Wilson, E. O. (2000). *Sociobiology: The new synthesis*. Harvard University Press. https://www.hup.harvard.edu/books/9780674002357

References

1. Green, A. (2020). Li Wenliang. *The Lancet, 395*(10225), 682. https://doi.org/10.1016/s0140-6736(20)30382-2

2. Kuo, L. (2020, April 10). Birth of a pandemic: Inside the first weeks of the coronavirus outbreak in Wuhan. *The Guardian*. https://www.theguardian.com/world/2020/apr/10/birth-of-a-pandemic-inside-the-first-weeks-of-the-coronavirus-outbreak-in-wuhan

3. World Health Organization. (2020, January 12). *Novel coronavirus – China*. https://www.who.int/emergencies/disease-outbreak-news/item/2020-DON233

4. He warned of coronavirus. Here's what he told us before he died. (2020, February 7). *The New York Times*. https://www.nytimes.com/2020/02/07/world/asia/Li-Wenliang-china-coronavirus.html

5. Qin. (2020, January 10). China reports first death from new virus. *The New York Times*. https://www.nytimes.com/2020/01/10/world/asia/china-virus-wuhan-death.html

6. Zheng, J. (2020). SARS-CoV-2: An emerging coronavirus that causes a global threat. *International Journal of Biological Sciences, 16*(10), 1678. https://doi.org/10.7150/IJBS.45053

7. Wipatayotin, A. (2020, January 22). First Thai infected with coronavirus. *Bangkok Post*. https://www.bangkokpost.com/thailand/general/1841199/first-thai-infected-with-coronavirus

8. Spiteri, G., Fielding, J., Diercke, M., Campese, C., Enouf, V., Gaymard, A., Bella, A., Sognamiglio, P., Moros, M. J. S., Riutort, A. N., Demina, Y. v., Mahieu, R., Broas, M., Bengnér, M., Buda, S., Schilling, J., Filleul, L., Lepoutre, A., Saura, C., ... Ciancio, B. C. (2020). First cases of coronavirus disease 2019 (COVID-19) in the WHO European Region, 24 January to 21 February 2020. *Eurosurveillance, 25*(9), 2000178. https://www.eurosurveillance.org/content/10.2807/1560-7917.ES.2020.25.9.2000178

9. Kretchmer, H. (2020, April 22). *Key milestones in the spread of the coronavirus pandemic*. World Economic Forum. https://www.weforum.org/agenda/2020/04/coronavirus-spread-covid19-pandemic-timeline-milestones/

10. Besser, L. (2020, March 18). Italy's coronavirus disaster: At first, officials urged people to go out for an aperitif. Now, doctors must choose who dies. *ABC News*. https://www.abc.net.au/news/2020-03-18/what-went-wrong-with-how-italy-handled-coronavirus/12062242

11. Italy imposes unprecedented lockdown to fight coronavirus: "We have run out of time." (2020, March 10). *CBS News*. https://www.cbsnews.com/news/coronavirus-italy-quarantine-lockdown-prime-minister-we-have-run-out-of-time/

12. Guerzoni, M. (2020, March 9). Coronavirus, Conte: «Non c'è più tempo, restiamo in casa». Il passo del premier dopo la chiamata al Colle. *Corriere Della Sera*. https://www.corriere.it/politica/20_marzo_09/coronavirus-conte-non-c-piu-tempo-restiamo-casa-passo-premier-la-chiamata-colle-3d15b7dc-6252-11ea-9897-5c6f48cf812d.shtml

13. World Economic Forum. (2020, March 21). *This chart shows how airlines are being grounded by COVID-19*. https://www.weforum.org/agenda/2020/03/this-chart-shows-how-airlines-are-being-grounded-by-covid-19/

14. Morrison, S. (2020, March 18). *Press Conference - Australian Parliament House, ACT*. PM Transcripts. https://pmtranscripts.pmc.gov.au/release/transcript-42738

15. Grupe, D. W., & Nitschke, J. B. (2013). Uncertainty and anticipation in anxiety: An integrated neurobiological and psychological perspective. *Nature Reviews Neuroscience, 14*(7), 488–501. https://doi.org/10.1038/nrn3524

16. Erickson, H. W. H. (2001). *The Report of Governor Bill Owens' Columbine Review Commission*. https://schoolshooters.info/sites/default/files/Columbine%20-%20Governor's%20Commission%20Report.pdf

17. Crocq, M.-A., & Crocq, L. (2000). From shell shock and war neurosis to posttraumatic stress disorder: A history of psychotraumatology. *Dialogues in Clinical Neuroscience, 2*(1), 47–55. https://doi.org/10.31887/DCNS.2000.2.1/macrocq

18. Grillon, C., Pine, D. S., Lissek, S., Rabin, S., Bonne, O., & Vythilingam, M. (2009). Increased anxiety during anticipation of unpredictable aversive stimuli in Posttraumatic Stress Disorder but not in Generalized Anxiety Disorder. *Biological Psychiatry, 66*(1), 47–53. https://doi.org/10.1016/j.biopsych.2008.12.028

19. Alvarez, R. P., Chen, G., Bodurka, J., Kaplan, R., & Grillon, C. (2011). Phasic and sustained fear in humans elicits distinct patterns of brain activity. *NeuroImage, 55*(1), 389–400. https://doi.org/10.1016/j.neuroimage.2010.11.057

20. Davis, M., Walker, D. L., Miles, L., & Grillon, C. (2010). Phasic vs sustained fear in rats and humans: Role of the extended amygdala in fear vs anxiety. *Neuropsychopharmacology, 35*(1), 105–135. https://doi.org/10.1038/npp.2009.109

21. Lebow, M. A., & Chen, A. (2016). Overshadowed by the amygdala: The bed nucleus of the stria terminalis emerges as key to psychiatric disorders. *Molecular Psychiatry, 21*(4), 450–463. https://doi.org/10.1038/mp.2016.1

22. Sullivan, G. M., Apergis, J., Bush, D. E. A., Johnson, L. R., Hou, M., & Ledoux, J. E. (2004). Lesions in the bed nucleus of the stria terminalis disrupt corticosterone and freezing responses elicited by a contextual but not by a specific cue-conditioned fear stimulus. *Neuroscience, 128*(1), 7–14. https://doi.org/10.1016/j.neuroscience.2004.06.015

23. Panksepp, J. (1998). *Affective neuroscience: The foundations of human and animal emotions.* Oxford University Press. https://global.oup.com/academic/product/affective-neuroscience-9780195178050

24. Grillon, C., Morgan, C. A., Davis, M., & Southwick, S. M. (1998). Effects of experimental context and explicit threat cues on acoustic startle in Vietnam veterans with posttraumatic stress disorder. *Biological Psychiatry, 44*(10), 1027–1036. https://doi.org/10.1016/S0006-3223(98)00034-1

25. Grillon, C., Morgan III, C. A., Davis, M., & Southwick, S. M. (1998). Effect of darkness on acoustic startle in Vietnam veterans with PTSD. *American Journal of Psychiatry, 155*(6), 812–817. https://ajp.psychiatryonline.org/doi/full/10.1176/ajp.155.6.812

26. Warda, G., & Bryant, R. A. (1998). Cognitive bias in acute stress disorder. *Behaviour Research and Therapy, 36*(12), 1177–1183. https://doi.org/10.1016/S0005-7967(98)00071-0

27. Mendl, M., Burman, O. H. P., Parker, R. M. A., & Paul, E. S. (2009). Cognitive bias as an indicator of animal emotion and welfare: Emerging evidence and underlying mechanisms. *Applied Animal Behaviour Science, 118*(3–4), 161–181. https://doi.org/10.1016/j.applanim.2009.02.023

28. Burman, O. H. P., Parker, R., Paul, E. S., & Mendl, M. (2008). A spatial judgement task to determine background emotional state in laboratory rats, *Rattus norvegicus*. *Animal Behaviour, 76*(3), 801–809. https://doi.org/10.1016/j.anbehav.2008.02.014

29. Clegg, I. (2018). Cognitive bias in zoo animals: An optimistic outlook for welfare assessment. *Animals, 8*(7), 104. https://doi.org/10.3390/ani8070104

30. The National Commission on Terrorist Attacks Upon the United States. (2004). *The 9/11 Commission Report.* https://9-11commission.gov/report/

31. Jaworski, A., Fitzgerald, R., & Constantinou, O. (2005). Busy saying nothing new: Live silence in TV reporting of 9/11. *Multilingua: Journal of Cross-Cultural and Interlanguage Communication, 24*(1–2), 121–144. https://doi.org/10.1515/mult.24.1-2.121

32. Battaglio, S. (2021, September 10). How 9/11 changed the way we watch TV news. *Los Angeles Times.* https://www.latimes.com/entertainment-arts/business/story/2021-09-10/9-11-tv-news-cable-change-how-we-watch

33. History.com Editors. (2010, August 13). *Ground Zero.* A&E Television Networks. https://www.history.com/topics/21st-century/ground-zero

34. History.com Editors. (2019, September 11). *September 11: Photos of the worst terrorist attack on U.S. soil.* A&E Television Networks. https://www.history.com/news/september-11-attacks-photos

35. Krause, J., & Ruxton, G. D. (2002). *Living in groups.* Oxford University Press.

36. Majolo, B., & Huang, P. (2018). Group living. In *Encyclopedia of animal cognition and behavior* (pp. 1–12). Springer International Publishing. https://doi.org/10.1007/978-3-319-47829-6_1865-1

37. Markham, A. C., & Gesquiere, L. R. (2017). Costs and benefits of group living in primates: An energetic perspective. *Philosophical Transactions of the Royal Society B: Biological Sciences, 372*(1727), 20160239. https://doi.org/10.1098/rstb.2016.0239

38. Majolo, B., de Bortoli Vizioli, A., & Schino, G. (2008). Costs and benefits of group living in primates: Group size effects on behaviour and demography. *Animal Behaviour, 76*(4), 1235–1247. https://doi.org/10.1016/j.anbehav.2008.06.008

39. Lewontin, R. C. (1970). The units of selection. *Annual Review of Ecology and Systematics, 1*(1), 1–18. https://www.annualreviews.org/doi/abs/10.1146/annurev.es.01.110170.000245?journalCode=ecolsys.1

40. Spencer, H. (1864). *The principles of biology* (Vol. 1). Williams and Norgate. https://www.gutenberg.org/files/54612/54612-h/54612-h.htm

41. Darwin, C. (1869). *On the origin of species by means of natural selection, or the preservation of favoured races in the struggle for life* (5th ed.). John Murray. https://darwin-online.org.uk/content/frameset?itemID=F387&viewtype=text&pageseq=1

42. Darwin, C. (1859). *On the origin of species by means of natural selection, or the preservation of favoured races in the struggle for life.* London: John Murray. https://darwin-online.org.uk/content/frameset?itemID=F373&viewtype=text&pageseq=1

43. Smith, J. M. (1964). Group selection and kin selection. *Nature, 201*(4924), 1145–1147. https://doi.org/10.1038/2011145a0

44. Hamilton, W. D. (1964). The genetical evolution of social behaviour. I. *Journal of Theoretical Biology, 7*(1), 1–16. https://doi.org/10.1016/0022-5193(64)90038-4

45. Darwin, C. (1871). *The descent of man* (Vol. 1). London: John Murray. https://darwin-online.org.uk/EditorialIntroductions/Freeman_TheDescentofMan.html

46. Trivers, R. L. (1971). The evolution of reciprocal altruism. *The Quarterly Review of Biology, 46*(1), 35–57. https://doi.org/10.1086/406755

47. Wilson, D. S., & Wilson, E. O. (2007). Rethinking the theoretical foundation of sociobiology. *The Quarterly Review of Biology, 82*(4), 327–348. https://doi.org/10.1086/522809

48. Ruse, M. (1980). Charles Darwin and group selection. *Annals of Science, 37*(6), 615–630. https://doi.org/10.1080/00033798000200421

49. Blomberg, S. P., & Garland, T. (2002). Tempo and mode in evolution: Phylogenetic inertia, adaptation and comparative methods. *Journal of Evolutionary Biology, 15*(6), 899–910. https://doi.org/10.1046/j.1420-9101.2002.00472.x

50. Wilson, E. O. (2000). *Sociobiology: The new synthesis.* Harvard University Press. https://www.hup.harvard.edu/books/9780674002357

51. Knafo-Noam, A., Uzefovsky, F., Israel, S., Davidov, M., & Zahn-Waxler, C. (2015). The prosocial personality and its facets: Genetic and environmental architecture of mother-reported behavior of 7-year-old twins. *Frontiers in Psychology, 6.* https://doi.org/10.3389/fpsyg.2015.00112

52. Henrich, J. (2004). Cultural group selection, coevolutionary processes and large-scale cooperation. *Journal of Economic Behavior & Organization, 53*(1), 3–35. https://doi.org/10.1016/S0167-2681(03)00094-5

53. Hamlin, J. K., Wynn, K., Bloom, P., & Mahajan, N. (2011). How infants and toddlers react to antisocial others. *Proceedings of the National Academy of Sciences, 108*(50), 19931–19936. https://doi.org/10.1073/pnas.1110306108

54. Rand, D. G., Greene, J. D., & Nowak, M. A. (2012). Spontaneous giving and calculated greed. *Nature, 489*(7416), 427–430. https://doi.org/10.1038/nature11467

55. Patterson, N., Richter, D. J., Gnerre, S., Lander, E. S., & Reich, D. (2006). Genetic evidence for complex speciation of humans and chimpanzees. *Nature, 441*(7097), 1103–1108. https://doi.org/10.1038/nature04789

56. Almécija, S., Hammond, A. S., Thompson, N. E., Pugh, K. D., Moyà-Solà, S., & Alba, D. M. (2021). Fossil apes and human evolution. *Science, 372*(6542). https://doi.org/10.1126/science.abb4363

57. Begun, D. R. (2019). *The real planet of the apes: A new story of human origins.* Princeton University Press. https://press.princeton.edu/books/paperback/9780691182803/the-real-planet-of-the-apes

58. Westerhold, T., Bickert, T., & Röhl, U. (2005). Middle to late Miocene oxygen isotope stratigraphy of ODP site 1085 (SE Atlantic): New constrains on Miocene climate variability and sea-level fluctuations. *Palaeogeography, Palaeoclimatology, Palaeoecology, 217*(3–4), 205–222. https://doi.org/10.1016/j.palaeo.2004.12.001

59. Andrews, P. (2019). Last common ancestor of apes and humans: Morphology and environment. *Folia Primatologica, 91*(2), 122–148. https://doi.org/10.1159/000501557

60. Wichura, H., Jacobs, L. L., Lin, A., Polcyn, M. J., Manthi, F. K., Winkler, D. A., Strecker, M. R., & Clemens, M. (2015). A 17-My-old whale constrains onset of uplift and climate change in east Africa. *Proceedings of the National Academy of Sciences, 112*(13), 3910–3915. https://doi.org/10.1073/pnas.1421502112

61. Andrews, P., & van Couvering, J. A. H. (1975). Paleoenvironments in the East African Miocene. In *Approaches to primate paleobiology* (Vol. 5, pp. 62–103). https://karger.com/books/book/4309/Approaches-to-Primate-Paleobiology

62. Wichura, H., Bousquet, R., Oberhänsli, R., Strecker, M. R., & Trauth, M. H. (2010). Evidence for middle Miocene uplift of the East African Plateau. *Geology, 38*(6), 543–546. https://doi.org/10.1130/G31022.1

63. Tomasello, M., Carpenter, M., Call, J., Behne, T., & Moll, H. (2005). Understanding and sharing intentions: The origins of cultural cognition. *Behavioral and Brain Sciences, 28*(5), 675–691. https://doi.org/10.1017/S0140525X05000129

64. Gervais, M., & Wilson, D. S. (2005). The evolution and functions of laughter and humor: A synthetic approach. *The Quarterly Review of Biology, 80*(4), 395–430. https://doi.org/10.1086/498281

65. Dunbar, R. I. M. (2022). Laughter and its role in the evolution of human social bonding. *Philosophical Transactions of the Royal Society B: Biological Sciences, 377*(1863). https://doi.org/10.1098/rstb.2021.0176

66. Committee on World Food Security. (2009). *Reform of the Committee on World Food Security.* https://www.fao.org/3/k7197e/k7197e.pdf

67. Terpou, B. A., Densmore, M., Thome, J., Frewen, P., McKinnon, M. C., & Lanius, R. A. (2019). The innate alarm system and subliminal threat presentation in posttraumatic stress disorder: Neuroimaging of the midbrain and cerebellum. *Chronic Stress, 3*, 247054701882149. https://doi.org/10.1177/2470547018821496

CHAPTER 3

Riders on the Storm: Chaos

I.

I lifted my chin. The skin on my neck was sticky. With all the hatches shut, the morning heat had converted the boat's cabin into a sauna. I stood up from the sofa, where I had fallen asleep in the wee hours, and pushed open the hatch. My eyes encountered a crisp blue sky. A relieving breeze kissed my cheek.

Falling onto the couch again, I stared through the hatch's square opening. Apart from the towering aluminum mast, all I could see was an azure canvas with some faint brush strokes of white. Occasionally the painting was visited by a soaring mutton bird. I continued to gaze for what could have been several minutes. *So, this is what calm is supposed to feel like.*

My thoughts glided over the last few days. After the airline canceled our flight to LA on Thursday, I searched for other flights to Mexico, but no one was flying. I began to concoct a backup plan to get us out of Brisbane. The thought of staying trapped inside the double-brick walls of our apartment, watching the world implode on TV, made me feel like a fox caught in a snare. We had to get out.

It was simple, I assured my husband, when I unveiled my plan on Thursday night. Even if the Australian borders were shut, which was looking more and more likely, we still had a reliable car and some time off work. We had driven 1,700 km north to Cairns for our wedding, so reaching Airlie Beach at just over 1,000 km wouldn't be difficult. From Airlie Beach, we would charter a sailboat and spend 7 days on our own, exploring the warm waters of the Whitsundays. At the time, I really believed that if we could just get away from everyone else, maybe we could ride out this episode of turmoil.

"The only thing is," I remember saying to him with some hesitation, "we need to leave tomorrow."

"Tomorrow?!?" he said.

I nodded and watched him spiral into panic.

"We can't go tomorrow," he protested. "I have too many things to pack. I'll need to get together my medical kit and service my car. Plus, we haven't ever been sailing on our own. I don't even know what we need." I could just see the checklists upon checklists of survival gear being compiled in his head. He looked horrified. "Can't we just go one day later?"

"We can't go any later," I said. "We have to take the boat for 7 nights starting Sunday. Unless you want to drive 12 hours in one day, we're going to have to break it up over two days…which means we leave tomorrow."

For the next few minutes, we both faced forward, frozen. My husband and I live and breathe travel. Planning the logistics of a trip is exhilarating, but having less than 24 hours to do it was a new challenge, even for us.

At last, I broke the silence.

"Here's the thing," I said. This was the phrase that my husband had come to love and hate about me—those quintessentially American words of bull-headed determination.

"All we have to do is drive there, which we know how to do, and get on the boat," I said.

We both paused to decide if we really believed the words I had just uttered.

"Once we're on the boat," I said, "we can figure out the rest."

It was nearing dusk by the time we set out on Friday. We dropped Leo at the sitter. We drove north until the sun slipped away, until the bright highway lights of the city faded behind us. And then we were alone, following the stars.

***

I glimpsed my phone—7:23 a.m. The charter company's morning roll call was at 8:00 a.m. It would be on VHF channel 82 or 74, I wasn't sure which. I wondered what I would report when my turn came. I hadn't really looked at the anchorage guide or figured out where we were going next. I seized the rolled chart above the sofa and tucked it under my arm. Then I tiptoed up the steps and slid back the companionway hatch.

The boat could be confused for a fixed dock, with the softly undulating waves barely nudging it. The breeze was now softer than my own breath. It's incredible how sensitive boats are to weather. Earlier that morning, when I came up to check out the noise, the boat was jumpy like a mother in an ambulance with her sick toddler. Now the boat was serene.

It reminded me of this contraption we saw once at the Museum of Old and New Art (MONA) in Hobart. MONA is an eerie place. You arrive by water, and then you're forced to climb 99 stairs between two stone walls, with your destination not even in sight. Once you're at the top, you see not just one museum, but an entire campus of bizarre, jagged structures. Most of MONA, however, is underground. Known for its confronting themes, MONA preserves in its hallways an impenetrable darkness that filled me with angst.

I sought refuge in the only window-rich building at MONA—a 1958 house designed by Roy Grounds, called the Round House. Coincidentally, it was also a library.

Inside the sunlit library was the most peaceful aura. The soothing sound of a pen traveling across coarse paper summoned me into the room. To my right, near one of the windows, was an instrument designed by artist Cameron Robbins. It entailed a funnel just outside the window, which used the power of the wind to drive a metal boom that passed through the window. Inside the window was a table bearing a series of horizontal spools. The wind-driven boom turned the first spool, which turned the others via a pulley of connecting wire. Attached to the end of the wire was a pen. The pen traced ellipses on slowly rolling paper. As the intensity and direction of the wind outside changed, so did the length and direction of the ellipses on the paper. The motion of the pen was hypnotic.

Even more captivating than the instrument itself was the long roll of paper, mounted on the library wall, which displayed months of the instrument's recordings. The pen's transcriptions of the wind's patterns captured everything from fair-day breezes to sudden gusts to dramatic storms. My eyes sought to find perfect circles within a sea of erratic lines and scratches, but I soon gave up. I stepped back to look at the whole picture. There was no real pattern to it. It was chaotic and magnificent at the same time. This was a truer representation of Earth, I thought, than one of those plastic globes sitting at the back of high-school classrooms. This depicted an Earth in continuous and ever-changing motion.

The intense morning sun illuminated my paper chart as I unrolled it on the cockpit table. I parked my phone on one corner and spread my hands across two other corners to stop them from curling. I located the spot where we were anchored, tucked inside the mainland in Funnel Bay. The chart showed shallow reefs on both sides of us. I peered over the port side of the boat, and sure enough, the water was a shallow turquoise. Backing out without running aground was going to be tricky.

I recalled the banging sounds of the previous night, and I began to worry. What if the wind had caused the boat to swing around the anchor, slamming us into the reef during the night? It did sound like something was bashing us. I grabbed my phone and Googled "Funnel Bay."

The first website I found said that prevailing weather in the Whitsunday region is east to southeast, with the strongest winds blowing at 15–20 knots from March to May. Another site said that Funnel Bay gets its name from the strong "bullets" of wind that zip down the hills of the high-sided valley into the bay. "Bullets" is a local name for sharp gusts created by the funneling effect that higher land masses have on the surface winds. These bullets can sometimes double the strength of the wind, catching vessels unawares, particularly in north-facing anchorages.[1]

Well, we have definitely been caught "unawares." I grasped my forehead and slid my hand over my eyes. I'd assumed that once we were anchored, we'd be safe. But what I hadn't factored in was the swing of the boat around the anchor. If the wind had shifted during the night, it could have blown us anywhere on the circle around the anchor, including right onto the reef.

*Skipper, how could you be so naï*ve? I learned an important sailing lesson that day: Boats don't sit still. No, the swing of a boat is no more predictable than the movement of a wind-driven pen.

II.

Nature often catches us unawares. Scientists used to believe the universe was orderly. In 350 B.C.E., Greek polymath Aristotle described the planets as celestial bodies with perfect and eternal motions.[2] The alleged perfection of these celestial bodies was reinforced by Nicolaus Copernicus in 1543 with his heliocentric theory of the universe.[3] In 1609, Johannes Kepler added that the planets move in ellipses rather than circles around the Sun,[4] but still, they were orderly ellipses.

Ideas about the order of the universe were swift to spread. Across the English Channel, Sir Isaac Newton expanded Kepler's work on planetary motion by describing laws pertaining to the motion of all objects.[5] Whether it was a planet in motion or an apple falling, Newton believed that all motion in nature followed the same rules.

These laws of motion put forward by these great thinkers of the Scientific Revolution gave rise to an orderly interpretation of the universe. If you knew all the properties of the system and you knew the laws of motion, then you could know exactly how the system would evolve over time. You could determine where something would travel far into the future. Following the Scientific Revolution, this philosophy of "determinism" became a dominating force in science. You probably recited Newton's laws in school: "For every action, there is an equal and opposite reaction," and so on. The universe, according to determinism, runs like a Swiss watch.

But not everyone was convinced. At the close of the 19th century, Oscar II, King of Sweden and Norway, was worried about the fate of our solar system. Yes, the planets were moving along elliptical orbits, he accepted, but who was to say that the solar system wouldn't eventually fall apart?[6]

Although the question seemed ludicrous to those under the spell of determinism, some of Europe's brightest thinkers quietly sympathized with Oscar II's doubts. Sir Isaac Newton himself observed, as the Greek astronomer Hipparchus had 18 centuries earlier, that the Moon's actual orbit suffered from many irregularities, making its location difficult to predict.[7] The problem frustrated Newton greatly, and he complained to fellow astronomer John Machin that his head never ached so much as when he was studying the Moon.[8] Newton suspected that the Moon's orbit was not only being affected by the gravitational pull of the Earth, but also by the gravitational pull of the Sun, which was warping the shape of its orbit.[7]

The problem that Newton encountered, whereby the Sun, the Earth, and the Moon all exert gravitational forces on each other, became known as the "three-body problem." If three bodies are moving in space and are mutually attracted to one another and we know their initial conditions, can we determine their subsequent motion?[6] The three-body problem seemed like a natural extension of the two-body problem, which Newton had solved using geometric deductions.[7] But the three-body problem was much more complex, and Newton's laws of motion were no match for it.[7]

Oscar II was about to turn 60, and this business of the Moon was really troubling his aging mind. Did the planets really just run around and around on closed, periodic orbits, or were their orbits changing over time, permitting the planets to one day run away?[9] Alas, the answer to this question lay in the elusive solution to the three-body problem. To find an answer, Oscar II sponsored an international competition for his 60th birthday. Prize competitions, generally sponsored by national academies, were a well-established way of seeking solutions to mathematical challenges. Oscar II promised a prize to any mathematician who could solve the three-body problem, accurately predicting where the Moon, Sun, and planets would go far into the future.[6]

Mathematicians across Europe got busy running calculations. Most of them took a simplified approach, approximating the three-body system using linear equations.[6] They bent the real, nonlinear system in their minds until it looked linear and could be solved. They would have justified their linearization of the problem by the assumption that, in the grand scheme of things, small variations in orbit would likely average out over time, resulting in a negligible impact on the order of the universe. But had they looked through a telescope at the real universe, they would have realized that they were wrong.

French mathematician Henri Poincaré took a different tack. While his colleagues were busy coming up with long, rigorous proofs that approximated the Moon's nature, Poincaré was interested in understanding the very "soul of the fact"—the way that nature really worked.[10] As a child, Poincaré had been fascinated with shapes, becoming enamored with the writings of French geometrician Michel Chasles.[10]

Poincaré's classmate recalled that Poincaré "saw no mathematical interest in drawing lines from memory."[10] Instead, he preferred to see nature as complex shapes, which he could describe "with a degree of perfection that the classical constructions did not attain."[10]

In the case of the three-body problem contest posed by Oscar II, rather than trying a linear approach, Poincaré chose to see a system of shapes. He used two balls to stand in for the Sun and the Earth, and a point (having very little mass relative to the others) to represent the Moon.[6]

If the Moon's orbit were perfect, Poincaré thought, it would end up at exactly the same spot, month after month. But in estimating the gravitational pull of the two balls on the point, Poincaré soon realized that the Moon did not end up at the same spot the next month. It ended up near it, but not on it. Poincaré drew a short line between the two spots to map the movement of the Moon from the first month to the second month. The next month, the Moon ended up somewhere else. He extended his curve to map the change from the second to the third month. He did this over and over, mapping the movement of the Moon from month to month. The curve of monthly change in the Moon's orbit, which resembled the outside of an oyster shell, resulted in what is now known as a Poincaré map.[11]

Poincaré's map of the Moon's motion still couldn't tell Oscar II exactly where the planets would end up far into the future, but it did answer his question, "Is the universe stable?"[6] The Poincaré map showed that although the position of the Moon varies from month to month, its path generally stays within certain limits. The outer bounds of the oyster shell, in other words, demonstrate that the Moon will not go flying off into space one day. Some sort of attractive force keeps pulling the Moon back.

Mathematician Ian Stewart provided a beautiful illustration of attractors in nature: If you drop a ping-pong ball over the ocean, it will fall, and, after a splash, it will float on the ocean's surface. But equally, if you hold it under the water and release it, it will pop up and eventually float on the ocean's surface. No matter where the ping-pong ball begins, even if it is thrashed about by the waves, it always ends up at its attractor—the ocean's surface. We cannot predict where the ping-pong ball will be in a minute from now, but we know that it will keep returning to its attractor. In other words, the ping-pong ball is part of a system that is stable overall, even if it's unstable from moment to moment.[12] Poincaré was the first to show that this is also the case with the Moon.

No other mathematician in Europe came close to providing a suitable answer to the three-body problem, so Oscar II awarded the prize to Poincaré.[6] Although he didn't have a clean mathematical proof involving simple equations, as many scientists would have liked, by proposing the concept of an attractor, Poincaré confirmed that the universe is not completely random and that the planets in our solar system are not likely to run away on us. That being said, his messy diagram of the Moon's motion from month to month indicated that nature is far from orderly.

Poincaré's winning entry was going to be published in the October 1889 volume of *Acta Mathematica*, where the competition had first been announced. But in July, there was a hiccup. Poincaré came across an error in his submission, one which he worried might invalidate his entire theory.[6] In his initial submission, Poincaré had assumed that the cross-section of the Moon's orbits, although it resembled a complicated oyster shell, would eventually form a closed loop. A closed loop would have meant that after a certain period, the Moon's motion would repeat itself.[6]

When Poincaré went over his math again, however, he came to a different conclusion. He realized that the Moon's orbit does not form a closed loop. In fact, the series of orbits *never* starts over again. There are times when the Moon will cross one of its previous paths momentarily, but it will never take the same orbit twice.[6] By spotting his error, Poincaré discovered that although the motion of the Moon is not random (it will not go flying off into space), it is also not predictable, because it doesn't repeat itself. Before Poincaré, the concept that something could move non-randomly but unpredictably had never been described. It would become the foundation of what is now known as chaos theory.[6]

Although this was a major discovery about the chaotic nature of the universe, Poincaré didn't take much notice of it at the time, because he was so consumed with another issue. When he ran his calculations again after correcting for his error, Poincaré was shocked to discover that the new map of the Moon's orbits looked entirely different from the first one he had made.[6] It was a completely different oyster. Most scientists at the time had assumed that if you only miscalculated the Moon's position slightly, it would only affect the subsequent trajectory of the Moon slightly. What Poincaré found, though, was that even small variations in the initial conditions of the problem tended to amplify over time, causing completely different trajectories for the Moon. This sensitivity to initial conditions was later called the "butterfly effect."[13]

Such an unpredictable and sensitive universe was not the orderly machine that the great thinkers of the Scientific Revolution had imagined. They assumed that nature would play by rules and be consistent to itself. They started by writing the rules and then tried to see if they could wrestle nature into fitting them. On the other hand, Poincaré started with no rules. He simply described what he saw. In the process, Poincaré discovered the inherent unpredictability in the behavior of complex natural systems—that is, chaos.[14] By doing so, he opened the first window into our chaotic world.

Since Poincaré's discovery, scientists have come to understand that much of nature is chaotic. Dripping faucets,[15] predator–prey ecologies,[16] measles epidemics,[17] the turbulent motion of cream added to coffee,[18] lightning bolts,[19] rain formation within clouds,[20] our own heartbeats[21]—all of these display chaotic behavior. These chaotic systems are sensitive to the tiniest changes, they appear random but they aren't, and (like a good serial killer) they have patterns, but they never repeat themselves too closely.[22]

Due to Poincaré's error, the winning entries to Oscar II's competition were published in *Acta Mathematica* almost a year later than planned. But by the time the *Acta* volume was circulated, people had long forgotten Poincaré's error and instead applauded his brilliance.[6] We now know that chaos in nature can be found nearly everywhere you look, but like Poincaré, you first have to be willing to look.

III.

If you are willing to look nature in the eye, without judgment or assumptions, she will reward you with the best of her secrets. Some of humanity's greatest scientists have discovered this. The difference between how Poincaré looked at the universe and how his contemporaries did comes down to a fundamental difference in how people apply Aristotle's scientific method—that is, how we *observe* what exists and *interpret* why it exists.[23] When we observe variations from the norm, do we interpret them as anomalies, as mere distractions from a mostly constant trend, or do we see them as significant in their own right? Asked differently, do we expect the world to be a rather orderly system, just ticking along, or a system that is chaotic?

In the fall of 1987, paleoceanographer Peter de Menocal embarked on a voyage across the Arabian Sea to probe the system.[24] A prominent professor of earth science, he was intent on uncovering how Earth's past climates might have shaped the lives of our ancestors. At the time of his voyage, the long-held view of human evolution was that a shift to a drier climate in Africa during the late Miocene (12–8 mya) led to a reduction in lush forests and the spread of dry grasslands.[25] The thought was that our early ancestors, who gradually developed the ability to walk upright and make tools, must have been better adapted to the new grassy savannas than were other apes at the time, who either went extinct or stayed in the forests that remained. Therefore, the adaptations that made us human arose from the environmental pressure of an expanding African savanna. This idea is known as the Savanna Hypothesis.[26] The Savanna Hypothesis is a rather tidy explanation—it shows the ascension of humans from other apes in a linear trajectory over time. But it seemed a far too orderly explanation of Earth's dynamics for De Menocal. So, he set sail in search of better observations.

If you dig deep enough, the Earth will reveal its past to you. Natural recordings—trees, lakes, glaciers, and sediments from the ocean floor—are all wrinkles on the Earth's face. They show the Earth's life story for millions of years before humans could ever record it. Seventeenth century Danish polymath, Nicolaus Steno, proposed that the Earth's rock layers (or strata) are laid down chronologically, one on top of the other.[27]

Layers of mud or sediment are deposited sequentially on the ocean floor, where they sit undisturbed for millennia.[28] Steno suggested that if you drill down into the Earth, each layer of rock that you see will represent a "slice" of time.

De Menocal was aboard a 143 m research vessel called *the JOIDES Resolution.* A few times each year, a new crew of 50-odd scientists from all over the world board the *JOIDES Resolution* as part of the International Ocean Discovery Program. Scientists in this program choose to head out to deep sea for months at a time, just to see what they can pull up from the ocean floor.[29] It was the Program's 117th leg, and De Menocal was one of 28 scientists aboard. Their goal was to drill at 12 locations around the northwest Arabian Sea.[24]

Proposed drilling locations were in the waters surrounding the Horn of Africa, not far from where humanity is thought to have started. The *JOIDES Resolution* team was in search of sediment cores that could give clues about environmental patterns at the dawn of humanity. Mud or sediment at the bottom of the ocean contains organic-rich matter, including microfossils, which can tell us about the oceanographic conditions that were present when the mud was deposited. Sedimentologists can analyze chemical components of the mud to decipher past changes in water temperature, nutrient composition, and movement of cold water from the deep over millions of years.[30] Sections of mud from areas like the Mediterranean Sea and the northwest Arabian Sea provide a high-resolution time series of ancient climates of East Africa, making them key targets for scientific ocean drilling.[24] These mud cores act as a pristine logbook for Earth's history.

As you can imagine, a logbook under the ocean is not particularly easy to access. This is why the *JOIDES Resolution* has several tonnes of machinery for drilling. After sailing for weeks or maybe even months, the crew arrives at their coordinates of interest. They lower a metal tube off the back of the ship and down through thousands of meters of water. The heavy metal tube then hits the bottom of the ocean. It punches deep into the layers of clay, like a hole punch going through a thick stack of papers. It captures a long core of sediments preserved in their original order of deposition. The crew then uses a crane to lift the core through the water and back onto the boat. Inside the tube is a long cylinder of Earth.[31]

When De Menocal heard "Core on deck!" squealed over the loudspeakers on the *JOIDES Resolution*, he grabbed his hard hat and rushed to the ship's laboratory to take a peek. Upon entering, he was hit with a chilly wall of air. The smell of putrefied mud flooded his sinuses. On the laboratory table lay open a 9 m core of deep-sea sediment which had been split down the middle lengthwise. Each half resembled a multilayered mud cake, with every layer a slightly different color to the next. De Menocal joined the other scientists who were gathered around beholding the rainbow of brown, green, yellow, and red.[32]

Having examined many sediment cores throughout his career, De Menocal had a sixth sense for the patterns they displayed. The yellow indicated sulfur content. The green indicated the presence of clay minerals like chlorite and glauconite. The red indicated presence of iron and an oxygenated environment.[33] But the predominant visual pattern on this core was one of alternating dark and light. De Menocal knew that these dark and light stripes provided just what he was looking for—an archive of Earth's past climates over the last 5 million years.

The dark stripes De Menocal saw were layers containing silt and dust. They were most likely deposited on the ocean floor when Africa had been cool and dry. When the Earth receives less radiation from the Sun, the Earth is colder, so more of the water is tied up in ice, which builds up in large glaciers near the Earth's poles.[34] During these glacial periods, less fresh water is available in rain clouds and in rivers and lakes, so many continents become drier. Dust and dirt are blown off the dry continent by prevailing winds and land on the ocean floor as a dark layer.[35]

The light stripes, on the other hand, were calcium carbonate—the same white, chalky mineral that gets stuck in your teeth when you chew an antacid. They likely corresponded to times when Africa had warmer, wetter conditions. In the open ocean, as much as half of the calcium carbonate comes from calcifying organisms no bigger than a pinhead, called forams. Forams build little shells of calcium carbonate around them.[36] When the Earth receives more radiation from the Sun and heats up, glaciers melt and rainfall becomes heavier. This creates a layer of fresh water on the surface of the ocean, while the bottom remains salty. Fresh water can hold more dissolved oxygen than salt water can, so the dissolved oxygen in the ocean rises to the top. Starved of oxygen near the ocean floor, the forams die en masse, leaving behind only a white carpet of their calcium carbonate shells.[37]

It was once thought that you could expect to see patterns of light and dark layers every 100,000 years or so, with the coming and going of the ice ages. But what De Menocal and the sedimentologists aboard the *JOIDES Resolution* witnessed did not correspond to these slow, 100,000-year cycles. What they saw was more like a bar code, with rapidly alternating layers of dark and light mud, indicating major climate changes from wet to dry about every 23,000 years.[32] These rapid changes in the deep-sea sediments hinted at a much more erratic environment for our ancestors than a one-time shift from lush forests to dry savanna. What could this mean ... if humanity was born not out of some definitive climate shift but out of environmental chaos?

IV.

At the same time that Peter de Menocal was observing chaos captured in Arabian Sea sediments, paleoanthropologist Rick Potts, founder of the Smithsonian's Human Origins Program, was observing chaos inland.

In the late 1980s, Potts was digging at Olorgesailie—a 1 million-year-old lake basin in the East African Rift—when he came across patterns in the mud that, at first, seemed like anomalies. Today the Olorgesailie basin is completely dry, a bowl of white dust, but the sediment there preserves some of the most securely dated artifacts, fossils, and climate records of the last million years. This slice of time includes the years before and during the emergence of our species, *Homo sapiens*.[38]

For most of his career, Potts had believed in the Savanna Hypothesis. He, like other paleoanthropologists of the time, had assumed that the adaptations that made us human, like walking on two feet, making tools, and forming complex societies, were the product of natural selection imposed by a shift towards a drier savanna habitat in Africa.[38] But like De Menocal, Potts couldn't quite fit the anomalies he was observing in the sediment into this orderly explanation of human evolution.

As Potts examined the rock wall of Olorgesailie's basin, he noticed thin layers of dark and light stacked on top of each other like a perfect croissant. These rapidly alternating layers hinted at frequent changes in the climate around the lake over the last million years. From the layers in the wall, it looked to Potts as if the lake had appeared one moment when there was abundant rainfall, had disappeared the next, had been reshaped by a volcanic eruption, had re-emerged with fresh water influx, had evaporated in an instant, had returned, and so on and so on.[38] Potts found at least 16 major environmental shifts in the period between 1.2 million and 400,000 years ago. There were also flaked stone tools at Olorgesailie, indicating that our early ancestors were likely in the region throughout this period of frequent change.[39]

To understand how our ancestors coped with such frequent changes in their climate, Potts turned to other primates in the region, both living and extinct.[40] Giant baboons, *Theropithecus oswaldi,* also known as giant gelada monkeys, used to be the most abundant primates in the East African Rift, including Olorgesailie.[38] Analysis of fossilized teeth indicates they were grazing animals with large, powerful jaws for eating coarse grass.[40] Males are estimated to have weighed over 80 kg, which is about the size of a female gorilla.[38] Despite their enormous size, giant baboons were constrained by their limited palate. They ate only grass. And when grass became scarce during one of the many climate shifts that occurred in East Africa, it became difficult for giant baboons to survive. Potts reasoned that giant baboons would have had to compete with all the other grazing animals for food. There wasn't enough grass to go around. Ultimately, giant baboons went extinct, likely because of their narrow niche.[40]

Although *Theropithecus oswaldi* went extinct around 500,000 years ago, some baboons are still alive today. Potts looked to them to see what they had that the giant baboon lacked. A relative of giant baboon, the common olive baboon *Papio anubis,* now inhabits 25 countries in Africa. Olive baboons evolved in a much more variable environment than giant baboons. Consequently, olive baboons are not specialized grazers. They will eat just about anything, from blossoms to tubers to animals the size of a small gazelle to the crops of local farmers.[38] When the climate shifts suddenly and certain food sources disappear, olive baboons can adapt with greater ease than giant baboons ever could. Their niche is broader. They're more flexible animals.[40]

In his hand, Potts held the teeth of the extinct giant baboon, a specialist who was unable to survive the rapid swings in the East African climate. And in his mind, Potts slowly snapped together the puzzle pieces of a new hypothesis for the origin of human beings. He called it the Variability Selection Hypothesis.[40]

According to Potts, key events in human evolution were shaped not by any single shift from wet to dry or any single type of habitat, as suggested by the Savanna Hypothesis. Instead, our evolution was shaped by environmental variability itself. This variability was a product both of Earth's imperfections and its continuously changing nature.[40] Just as Poincaré found with the Moon, the motion of the Earth is far from orderly. The Earth is asymmetric, so it wobbles when it spins. Like a top that is slightly heavier on one side, the Earth spins chaotically, falling over and coming back up again. Moreover, the Earth's lava and tectonic plates are in continuous motion. The combination of these external and internal forcing mechanisms creates a complex series of nonlinear feedback loops that make for a chaotic world.[41] Potts reasoned that in a highly variable world, those animals who survive have likely evolved to have greater flexibility in their behavior. These flexible animals have a better chance of surviving in novel, unpredictable, or chaotic environments.

In light of the Variability Selection Hypothesis, it is no surprise that behavioral flexibility is the main characteristic that unites all modern great apes (gorillas, chimpanzees, orangutans, and humans).[42] Great apes are very mobile, and we live a long time, so why would we be suited to only one sort of environment? Wouldn't it make more sense for us to be able to fine-tune our behavior to a variety of environments, in case they change?[43]

One of our oldest ancestors to have diverged from other great apes is a prime example of behavioral flexibility. Meet Ardi, a nickname given by scientists to *Ardipithecus ramidus*. Ardi is thought to have lived 4.4 mya in the Middle Awash and Gona areas of Ethiopia, East Africa. Her bones were discovered in 1994 alongside animal remains. The enamel on her teeth is neither very thick nor very thin, suggesting she ate a variety of foods and was not confined to fruit or grasses. Although she is thought to have lived in a wooded environment, Ardi's skeleton is different to modern forest-dwelling chimpanzees. She has a divergent large toe combined with a rigid foot, better suited to standing than grasping. Her pelvis, which was reconstructed from a crushed specimen, shows adaptations that combine tree climbing and walking.[44] These findings indicate that although members of *Ardipithecus ramidus* could likely walk upright, they also kept their ability to climb trees. They had, as Potts would have expected, behavioral flexibility.

Ardi's ability to walk upright in a wooded environment contradicts the Savanna Hypothesis for the origin of bipedalism, which insists that our ancestors began to walk on two feet as the climate dried and our environment became more open and grassier. Ardi might have descended from an earlier species of *Ardipithecus* that has been found in the same area of Ethiopia, *Ardipithecus kadabba*. A toe bone from this species, dated as 5.2 million years old, already indicates features of bipedal walking. Animal remains from the site suggest that the hominins who lived there inhabited a mixture of woodlands and grasslands, and that they had plenty of access to water via lakes and springs.[45] They lived in a variable environment, so it makes sense that they were well adapted to all kinds of conditions.

Considering the evidence of Earth's variable climates and the flexible adaptations of *Ardipithecus*, it is doubtful that our ancestors would have ever experienced a single shift from forest to savanna, as proposed by the Savanna Hypothesis. Instead, their environment was likely filled with continuous swings between wetter and drier times, corresponding to the Earth's changing motion. These swings altered their resources and changed their landscapes,[46] affecting what they could eat, where they could establish shelter, when they could have offspring, and where they could move. Like the other animals living in East Africa, our ancestors were at the mercy of the rain, which was at the mercy of the Earth's motion. Only animals who were sufficiently flexible in their behavior would have been able to survive the chaos.

We once thought that chaos in nature was the exception. Like a boat that sits perfectly still, we thought that nature should behave according to certain rules. Prominent leaders of the Scientific Revolution believed that the world worked in a tidy way, and we humans could have certainty about it. But what Henri Poincaré, Peter de Menocal, Rick Potts, and many other scientists have since discovered is that chaos isn't the exception to the rule—it is the rule. And if, from the very beginning, our environment was characterized by chaos, then humans must have somehow evolved to be very good at handling it.

CHAPTER 3 REFERENCES

Free Resources

Borwein, J., & Rose, M. (2012, November 19). Explainer: What is chaos theory? *The Conversation.* https://theconversation.com/explainer-what-is-chaos-theory-10620

Further Reading

Gleick, J. (1987). *Chaos: Making a new science.* Penguin Books. https://www.penguin.com.au/books/chaos-9781409027706

References

1. Maritime Safety Queensland. (2018). *Port Procedures and Information for Shipping: Whitsundays. 6. Weather Information.* https://www.publications.qld.gov.au/dataset/90530ffc-f6c5-459c-89f1-a5f9c9cfb95b/resource/366b587a-61e9-4d56-8639-65da7a48d1af/download/whitsundays_ppm_master_march_2023-sec-6.pdf

2. Aristotle. (350 B.C.E.). *On the heavens* (Translated by J. L. Stocks). http://classics.mit.edu/Aristotle/heavens.html

3. Copernicus, N. (1543). *De revolutionibus orbium coelestium [On the revolutions of the heavenly spheres].* Johannes Petreius. https://www.geo.utexas.edu/courses/302d/Fall_2011/Full%20text%20-%20Nicholas%20Copernicus,%20_De%20Revolutionibus%20(On%20the%20Revolutions),_%201.pdf

4. Kepler, J. (1609). *Astronomia nova [aitiologetos (romanized form)].* G. Voegelinus. https://library.si.edu/digital-library/book/astronomianovaa00kepl

5. Newton, I. (1686). *Philosophiae naturalis principia mathematica [Mathematical principles of natural philosophy].* Jussu Societatis Regiae ac typis Josephi Streater. Prostat apud plures bibliopolas. https://digital.library.sydney.edu.au/nodes/view/7454

6. Barrow-Green, J. (1994). Oscar II's prize competition and the error in Poincaré's memoir on the three body problem. *Archive for History of Exact Sciences, 48*(2), 107–131. https://www.jstor.org/stable/41133989

7. Cook, A. (2000). Success and failure in Newton's lunar theory. *Astronomy and Geophysics, 41*(6), 6.21-6.25. https://doi.org/10.1046/j.1468-4004.2000.41621.x

8. Brewster, D. (2010, first published in 1855). *Memoirs of the life, writings, and discoveries of Sir Isaac Newton* (Vol. 2). Cambridge University Press. https://doi.org/10.1017/CBO9780511792687

9. Holmes, P. (1990). Poincaré, celestial mechanics, dynamical-systems theory and "chaos." *Physics Reports, 193*(3), 137–163. https://doi.org/10.1016/0370-1573(90)90012-Q

10. Gray, J. (2013). *Henri Poincaré*: A scientific biography. Princeton University Press. https://press.princeton.edu/books/hardcover/9780691152714/henri-poincare

11. Gleick, J. (1987). *Chaos: Making a new science*. Penguin Books. https://www.penguin.com.au/books/chaos-9781409027706

12. Stewart, I. (1999). Symmetry-breaking cascades and the dynamics of morphogenesis and behaviour. *Science Progress, 82*(1), 9–48. https://doi.org/10.1177/003685049908200102

13. Abraham, R., & Ueda, Y. (2001). *The chaos avant-garde: Memories of the early days of chaos theory* (Vol. 39). World Scientific. https://doi.org/10.1142/4510

14. *Chaos*. (2023). Merriam-Webster.Com. https://www.merriam-webster.com/dictionary/chaos

15. Coullet, P., Mahadevan, L., & Riera, C. S. (2005). Hydrodynamical models for the chaotic dripping faucet. *Journal of Fluid Mechanics, 526*, 1–17. https://doi.org/10.1017/S0022112004002307

16. Hastings, A., Hom, C. L., Ellner, S., Turchin, P., & Godfray, H. C. J. (1993). Chaos in ecology: Is mother nature a strange attractor? *Annual Review of Ecology and Systematics, 24*(1), 1–33. https://doi.org/10.1146/annurev.es.24.110193.000245

17. Dalziel, B. D., Bjørnstad, O. N., van Panhuis, W. G., Burke, D. S., Metcalf, C. J. E., & Grenfell, B. T. (2016). Persistent chaos of measles epidemics in the prevaccination United States caused by a small change in seasonal transmission patterns. *PLOS Computational Biology, 12*(2), e1004655. https://doi.org/10.1371/journal.pcbi.1004655

18. Knight, J. (2002). Honey, I shrunk the lab. *Nature, 418*(6897), 474–475. https://doi.org/10.1038/418474a

19. Gomes, C., Cooray, V., Fernando, M., Montano, R., & Sonnadara, U. (2004). Characteristics of chaotic pulse trains generated by lightning flashes. *Journal of Atmospheric and Solar-Terrestrial Physics, 66*(18), 1733–1743. https://doi.org/10.1016/j.jastp.2004.07.036

20. Tölle, M. H., & Krueger, S. K. (2014). Effects of entrainment and mixing on droplet size distributions in warm cumulus clouds. *Journal of Advances in Modeling Earth Systems, 6*(2), 281–299. https://doi.org/10.1002/2012MS000209

21. Skinner, J. E., Goldberger, A. L., Mayer-Kress, G., & Ideker, R. E. (1990). Chaos in the heart: Implications for clinical cardiology. *Nature Biotechnology, 8*(11), 1018–1024. https://doi.org/10.1038/nbt1190-1018

22. Bishop, R. (2017). Chaos. *The Stanford Encyclopedia of Philosophy.* https://plato.stanford.edu/entries/chaos/

23. Ross, S. (1962). Scientist: The story of a word. *Annals of Science, 18*(2), 65–85. https://doi.org/10.1080/00033796200202722

24. Prell, W. L., Niitsuma, N., Emeis, K.-C., Rabinovitz, P. D., Myer, A. W., & Garrison, L. E. (1987). *Leg 117: Preliminary report.* http://www-odp.tamu.edu/publications/prelim/117PREL.PDF

25. Doman, J. H., & Early, E. G. (2022). Late Miocene and earliest Pliocene paleoecology of Africa. In *African paleoecology and human evolution* (pp. 24–32). Cambridge University Press. https://doi.org/10.1017/9781139696470.004

26. Domínguez-Rodrigo, M. (2014). Is the "savanna hypothesis" a dead concept for explaining the emergence of the earliest hominins? *Current Anthropology, 55*(1), 59–81. https://doi.org/10.1086/674530

27. Vai, G. B. (2021). Leonardo da Vinci's and Nicolaus Steno's geology. *Earth Sciences History, 40*(2), 293–331. https://doi.org/10.17704/1944-6187-40.2.293

28. Mulder, T., Hüneke, H., & van Loon, A. J. (2011). Progress in deep-sea sedimentology. In *Developments in sedimentology* (Vol. 63, pp. 1–24). Elsevier B.V. https://doi.org/10.1016/B978-0-444-53000-4.00001-9

29. International Ocean Discovery Program. (2023). *Exploring the Earth under the sea.* Scripps Institution of Oceanography. https://www.iodp.org/

30. Berger, W. H. (2011). Geologist at sea: Aspects of ocean history. *Annual Review of Marine Science, 3*(1), 1–34. https://doi.org/10.1146/annurev-marine-120709-142831

31. Walczak, M., Opdyke, B., & Post, A. (2017, May 30). *Going back in time...with mud.* Australian Academy of Science. https://www.science.org.au/curious/earth-environment/sediments

32. de Menocal, P. B. (2014). Climate shocks. *Scientific American, 311*(3), 48–53. https://doi.org/10.1038/scientificamerican0914-48

33. Hanano, D. (n.d.). *The many moods of mud.* Munsell Color. Retrieved September 13, 2023, from https://munsell.com/color-blog/the-many-moods-of-mud/

34. Hartmann, D. L. (2016). History and evolution of Earth's climate. In *Global physical climatology* (pp. 261–291). Elsevier. https://doi.org/10.1016/B978-0-12-328531-7.00009-8

35. Windom, H. L. (1975). Eolian contributions to marine sediments. *SEPM Journal of Sedimentary Research, 45.* https://doi.org/10.1306/212F6DAC-2B24-11D7-8648000102C1865D

36. Jacob, D. E., Wirth, R., Agbaje, O. B. A., Branson, O., & Eggins, S. M. (2017). Planktic foraminifera form their shells via metastable carbonate phases. *Nature Communications, 8*(1), 1265. https://doi.org/10.1038/s41467-017-00955-0

37. Kaiho, K. (1999). Effect of organic carbon flux and dissolved oxygen on the benthic foraminiferal oxygen index (BFOI). *Marine Micropaleontology, 37*(1), 67–76. https://doi.org/10.1016/S0377-8398(99)00008-0

38. Potts, R. (1996). *Humanity's descent: The consequences of ecological instability.* Avon Books. https://books.google.com.au/books/about/Humanity_s_Descent.html?hl=fr&id=CDPTOAAACAAJ&redir_esc=y

39. *Olorgesailie: Life and times of the handaxe makers.* (2022). Smithsonian Institution. https://humanorigins.si.edu/research/east-african-research-projects/olorgesailie-kenya

40. Scott, M. (2016, November). *Climate and human evolution.* Climate.Gov. https://www.climate.gov/news-features/climate-and/climate-and-human-evolution#:~:text=Variability%20selection%20proposes%20that%20major,any%20one%20environment%20or%20trend.

41. Campisano, C. J. (2012). Milankovitch cycles, paleoclimatic change, and hominin evolution. *Nature Education Knowledge, 4*(3), 5. https://www.nature.com/scitable/knowledge/library/milankovitch-cycles-paleoclimatic-change-and-hominin-evolution-68244581/

42. Malone, N., Fuentes, A., & White, F. J. (2012). Variation in the social systems of extant hominoids: Comparative insight into the social behavior of early hominins. *International Journal of Primatology, 33*(6), 1251–1277. https://doi.org/10.1007/s10764-012-9617-0

43. Belsky, J., & Pluess, M. (2013). Beyond risk, resilience, and dysregulation: Phenotypic plasticity and human development. *Development and Psychopathology, 25*(4pt2), 1243–1261. https://doi.org/10.1017/S095457941300059X

44. Smithsonian Institution. (2022). *Ardipithecus ramidus.* https://humanorigins.si.edu/evidence/human-fossils/species/ardipithecus-ramidus

45. Smithsonian Institution. (2022). *Ardipithecus kadabba.* https://humanorigins.si.edu/evidence/human-fossils/species/ardipithecus-kadabba

46. Antón, S. C., Potts, R., & Aiello, L. C. (2014). Evolution of early *Homo*: An integrated biological perspective. *Science, 345*(6192). https://doi.org/10.1126/science.1236828

CHAPTER 4

Lookin' Out My Back Door: Vigilance

I.

A snorkel-sized geyser sounded behind the boat as my husband's head broke the surface.

"How does it look?" I called.

He proceeded to plop one fin onto the teak-topped swim platform, followed shortly by the other. He pushed himself up and reached for his towel, proceeding to dry the beads of water from his tan skin.

"I swam all around and couldn't see anything," he said. "The reef doesn't start until over there." He pointed towards the shoreline.

"That's good," I said. "Was there much room under the keel?"

"I couldn't tell, but it wasn't sitting on the bottom. It's a sandy bottom anyway," he said.

I nodded. The boat had swung a lot throughout the night, but it looked as if it hadn't crashed into the reef. This was a huge relief ... at least temporarily. There were still six more nights to go. I didn't want to know how much this boat was worth. It was practically new. Apart from the spotless fiberglass hull and all the beautiful teak (which, for the cost, may as well be gold), there were so many moving parts on it—expensive "marine" parts. Already I regretted our stupid decision to pass up the charter company's offer to purchase additional insurance. It meant we would spend the whole week holding our breath, waiting for something to break.

My husband showered, and I finally washed the dishes and salsa-encrusted pan from the breakfast chilaquiles I had cooked. After stacking the dishes in the miniature dish rack, I returned to the cockpit to one of the two touchscreen chart plotters. As I turned it on, the chart of Funnel Bay appeared. I spread my fingers to zoom in on our location. It looked like there were shallow reefs on all three sides of us. We would have to reverse straight back until we reached an area wide enough to turn the boat around and leave the bay. It would take careful control at the helm as we lifted the anchor. I was more than happy to hand that task over to my husband.

There are plenty of enjoyable moments in sailing, but there are also moments when you watch your life savings swinging from the bow of a boat. The anchor emerged from the surface caked with mud. I knelt over the bow and stared down at it. Suspended in the air by its chain, the anchor began to swing, making random ellipses from side to side. The boat, freed of all attachments, began to drift aimlessly towards the shore. I shouted back at my husband to reverse. I heard the engine roar. With the boat kicked into reverse, the anchor shifted its motion to a back-and-forth swing. I squinted and braced myself for its inevitable crash into the bow. I clenched my teeth and slammed the button to raise it. A few long seconds later, I saw the anchor's shank appear over the bow roller. The arms soon followed, mud and all. I let out my breath. *No gashes for now.*

Once we had made it out of Funnel Bay, we were ready to set sail. We waited until we were well and truly past Pioneer Rocks, and then my husband turned us into the wind and slowed right down. The wind superseded the engine as the dominant sound. In all directions, the sky boasted puffy, white cumulus clouds.

I took over the helm while my husband walked up the side of the boat to the mast. I ensured we remained pointed into the wind as he unzipped the boom bag. He looked back at me, ready to raise the mainsail, and I looked at the wind vane. Once we were dead into the wind, I yelled out, "Okay, raise it." He began to pull down hard on the main halyard in body-sized yanks. The top of the mainsail soon appeared in view.

This boat had a lazy jack system, which is a triangle of ropes on either side of the mainsail designed to guide the mainsail into a nice pile over the boom when it comes down. In theory these ropes would also guide the mainsail up as you hoisted it, but as my husband continued to pull on the halyard, I saw the sail's battens catching on the ropes of the lazy jack. I yelled out to him to hold up. We needed to drop the sail a meter or so and start again.

You can see why many people just give up and get a powerboat. The lazy jack was more of an annoyance than an aid to the lazy, but even without its obstructions, I wondered how someone sailing alone would manage to be both at the mast hoisting the mainsail and at the helm keeping the boat pointed into the wind. *How does anyone sail alone?*

Most people don't sail alone, I reasoned. It's not just about needing to hoist and steer simultaneously. Even while underway, you need one person looking ahead to maintain the boat at a consistent angle to the wind while the other is looking up and trimming the sails. Not only that, it helps to have at least two or more sets of eyes scanning the horizon.

There are boats, Jet Skis, and snorkelers to watch out for, reefs and rocks to stay clear of, and strong gusts that can catch you by surprise. With just one person, you would be guessing about what's going on half the time. I suppose it also helps to have two people on board in case you crash and start to sink. You want someone getting the lifeboat ready while the other person shouts, "Mayday, Mayday, Mayday" into the radio.

I don't understand how people cross the ocean alone. A friend of a friend just did it for a short jaunt from French Polynesia. It was several days and nights of him alone at the helm. In theory, you can stop the boat in the middle of the ocean by "heaving to." But according to this guy, most solo sailors prefer to keep sailing, taking short naps to keep themselves from going delirious. If you have a working autopilot, you can set the boat on a bearing while you're asleep, but this won't prevent you from crashing if something happens to cross your course, like a cargo vessel, a whale, or a floating shipping container. No, alone at sea, you basically have to shut your eyes and pray.

The Whitsunday Islands provide plenty of shelter from the open ocean so, thankfully, all we needed to worry about were the tiny gusts coming across the Whitsunday Passage. Still, with our limited sailing experience and the lazy jack catching our mainsail, we struggled even with the gusts.

After a lot of toing and froing—"Hold up," ... "Drop it," ... "Okay, raise it,"—we eventually inched the mainsail up to where it nearly met the top of the mast. My husband returned to the cockpit to unfurl and trim the jib, which was a hundred times easier than the mainsail. Then he grabbed the mainsheet, which controls the angle of the mainsail relative to the boat, and he slowly let it out to the port side. As he did, I steered the boat away from the wind, towards the east. The Whitsunday breezes began to spill over the starboard lip of the bow, filling the two sails. As I turned more, the wind tipped the boat on its side by 10 degrees, the way a puff of breath pushes over the flame on a birthday candle.

We gained speed. The wind rushed across my right ear. It became tough for me to hold the helm on course—the boat wanted to resist. My husband's head tilted backward, watching the sails for luffing. When the jib sheet was too taught to pull any further by hand, he yelled out for the winch handle, which was in a holder next to my left calf. I handed it to him, keeping a tight hold of the helm with my right hand so that the nose of the boat wouldn't wander back into the wind.

He cranked the jib sheet as far as he could and tied it off. Behind his head, the teal water rushed by. We were heeled over 15 degrees and were zipping along. I saw my husband's shoulders relax for the first time in days. He pulled out his phone. A few seconds later, Jimmy Buffett's "Margaritaville" boomed through the cockpit speakers. My joy must have been automatic, because before I realized, the breeze was spilling into my cheeks through the corners of my smile. I felt a sense of triumph. Individually, we were clueless. But by working together, we had somehow managed to harness the wind.

II.

Sailing disaster movies would have you believe that sailing alone is one of the riskiest things a person could do. What they often don't show you, though, is that *technically* a sailor is never alone. At any given time, there is a global emergency system watching out for sailors in trouble.

The Emergency Position-Indicating Radio Beacon (EPIRB) is a device only slightly larger than a flashlight. All ocean-faring vessels are advised to carry at least one on board. The idea of an EPIRB is that if you find yourself in a life-threatening situation on the ocean, and Maydays, SOS, and visual signals have all failed to get a response, you can manually activate the EPIRB to transmit a distress signal to a passing satellite. Or, if your EPIRB sinks below 3 m of water, it will activate automatically, float to the surface, and start transmitting.[1]

Nowadays, if you activate an EPIRB, it will alert local search and rescue authorities no matter where you are in the world. But this wasn't always the case. Creating a global system of emergency assistance required a substantial dose of international cooperation, at a time when certain governments were already not getting along.

The global system originated in the early 1970s. Two U.S. congressmen were aboard a plane flying over Alaska when the plane went missing. Search and rescue efforts looked for the men, but no trace of them or the plane was ever found. After that, Congress mandated that all U.S. aircraft carry an Emergency Locator Transmitter, which would automatically activate after a crash and transmit a homing signal. The only problem was that Emergency Locator Transmitters relied on the old technology of radio transmissions at 121.5 MHz. These radio signals were often cluttered, and another aircraft had to be within range to receive the transmission.[2] This was not ideal when airplanes were, for instance, flying over the open ocean, far away from anyone else.

The U.S. National Aeronautics and Space Administration (NASA) realized that search and rescue efforts in remote areas could really benefit from satellite technology. After having put a man on the Moon in 1969, NASA's focus had shifted to using satellite technology for other purposes. They reasoned that a good system for remote search and rescue would require global coverage and would need to operate 24 hours a day, 7 days a week. But a global alert system aided by satellites would require cooperation with other nations who were launching satellites at the time, namely, Canada, France, and the Soviet Union.[2]

At the time, the Cold War dominated international relations, with most countries organized into Eastern and Western blocs. The countries had opposing ideologies, and cooperation between the blocs seemed inconceivable. There was one tiny commonality, however, in that countries on both sides were looking for ways to justify their investment in space technology to their citizens. Initially, NASA began working with Canada and France from the Western bloc to develop both the beacon and satellite technology for a system that would become known as Search and Rescue Satellite-Aided Tracking (SARSAT). They designed the beacons (later called EPIRBs) in such a way that each beacon was uniquely identified by its own digital signature. In addition, they gave the system its own dedicated digital frequency of 406 MHz. Meanwhile, in the Eastern bloc, the Soviet Union worked on developing their own satellite-aided search and rescue system, known as COSPAS.[2]

The political ice that separated East and West began to thaw with the success of the Apollo–Soyuz mission in 1975. American and Soviet mission commanders exchanged the first international handshake in space through an open hatch in the Soyuz capsule. A window for cooperation had opened. Policymakers on both sides of the Cold War saw an opportunity to show their citizens the peaceful outcomes of space technology.[3] An international satellite-assisted emergency response system seemed like a logical next step for peaceful cooperation. After a series of diplomatic talks, all four countries signed a Memorandum of Understanding on November 23, 1979. The Eastern and Western systems joined to form COSPAS–SARSAT.[2,3] Even the Soviet invasion of Afghanistan one month later, which outraged American and European governments, did not stop the international cooperation on COSPAS–SARSAT from going ahead.

The global COSPAS–SARSAT system came to its first rescue on October 11, 1982. Trimaran "Gonzo" was sailing in the Atlantic, 300 nautical miles east of Boston, when it was hit by a 9 m wave, causing it to capsize.[2,4] In the upside-down boat, the three-man crew donned their survival suits to prevent hypothermia, escaped through a hole they cut in the hull, and activated the boat's EPIRB. The signal was picked up by a Coast Guard plane passing overhead, but Gonzo's exact location could not be determined.[4] Shortly after, a COSPAS–SARSAT satellite passed overhead and received the transmitted signal, allowing authorities to pinpoint the exact location of the EPIRB. The U.S. Coast Guard was able to rush to the location and rescue all three sailors. The sailors were found holding onto lines that they had tied across the bottom of their overturned boat.[4] Although at this stage the COSPAS-SARSAT system was barely in the test phase, the rescue of the Gonzo crew showed that the system worked brilliantly. In 1984, COSPAS–SARSAT was declared fully operational.[2]

Today the COSPAS–SARSAT system continues to look out for aircraft and marine vessels in distress all over the world. The basic system remains the same, with 406 MHz being the digital frequency reserved for distress signals. Today there are 62 satellites orbiting the Earth listening for distress calls. The four founding nations have been joined by 45 other nations, user states, and organizations operating ground stations and mission control centers.[2] It continues to be one of the most successful examples of intergovernmental cooperation. These nations realized that for the purposes of keeping watch, it was better if they all worked together.

If you have a registered EPIRB, someone is watching out for you, no matter where you are in the world. When you're in trouble and send out an EPIRB distress signal, it's possible that a government other than your own—perhaps even one you'd considered an enemy—will come to your aid. With a total area of 361 *million* km^2, the ocean is way too vast for any one nation to patrol it. If it weren't for EPIRBs, sailors in distress would have to rely on the slim chance that a passing ship would come to their rescue. But because nations have banded together to develop a global system of satellites listening for distress calls, you can feel safe even when you're out in the middle of the ocean.

III.

Just as the human world watches out for stranded sailors, many other social animals work together to keep watch. A recent visit to Australia Zoo gave me an intricate picture of how this works.

Australia Zoo is just an hour north of Brisbane. You exit west from the motorway towards Beerburrum, and immediately you are immersed in the densely forested Steve Irwin Way. I pulled into the zoo's parking lot, turned off the Jimny, and donned my yellow backpack containing my notebook and pencil. I approached the ticket counter to purchase a ticket. It was early afternoon at the start of a hot Queensland summer, and no other visitors were in sight. Upon entering the zoo, I walked only a few steps before boarding an air-conditioned shuttle bus, which transported me, the only rider, to "Africa."

As soon as the shuttle approached, the heads of giraffes, chewing meditatively, and the distinctive markings of a grazing zebra signaled that I was, indeed, in "Africa." I stepped off the bus and was welcomed by tribal music emanating from the eaves of the gateway shelter. Following the pink stone pathway around to the left, I encountered two stoic rhino butts. Beyond them, in the distance, I glimpsed an enclosure that resembled a rocky Saharan environment. The exhibit was dusty and sun weary. From afar, I couldn't spot a single animal, but it looked like the kind of scene from which Lawrence of Arabia would emerge.

Out of nowhere, a petite furry back appeared. The animal proceeded to hobble along the elevated surface of a fallen log. The tiny, pointed snout was unmistakable. "There's one!" I whispered to myself with delight. A meerkat.

The meerkat scurried to the top of an African aloe tree, the highest point in the exhibit. The top of the aloe was flattened, as if from repetitive use. The meerkat perched upright, squatting on the tripod of her two hind legs and tail. Her front paws hung limply over her belly, like they were ready to play a piano sonata. Instantly, her black-socketed eyes got to work, surveying the zoo in every direction.

Just behind the exhibit, a 2 tonne white rhino ambled by. The meerkat seemed unbothered by the rhino, orienting instead to every other movement in her environment. Her head jerked like a lawn sprinkler, fine-tuning her orientation. On each side of her head perched a tiny black ear. I watched as she scanned the distance in all directions, vigilant to everything going on around her.

I observed the meerkat's observations. Below her and to her left, three of her flatmates were resting their heads on a rocky ledge. I watched them calmly look around their environment and overheard their soft chatter. No other humans were around the exhibit at this time of day. It was a hot and lazy sort of afternoon. One of the resting meerkats decided to flip onto her back and have a delicious stretch. At this point, I couldn't help but hear "Hakuna Matata" from *The Lion King* playing in my head—it never gets old. Perhaps it was because of the peaceful environment that the meerkat on watch soon felt comfortable enough to stand all the way up on her hind feet, elongating her ballerina-like legs and straightening her furry back. Balancing on her hind tippy toes with the assistance of her tail, she now had a much better vantage of the zoo.

She was displaying what meerkat biologists would call "sentinel behavior." Unlike Timon in *The Lion King*, meerkats in the wild are seldom found by themselves. They live in groups of up to 50 individuals.[5] You'll often see one meerkat on sentry duty, standing high on a mound. This sentinel meerkat is keeping an eye out for potential threats in the vicinity.[6] In the wild, meerkats live in flat, arid environments, leaving them few places to hide from predators like jackals, wildcats, snakes, and eagles.[7] To protect themselves, they must be constantly vigilant. But remaining vigilant and staying on guard is a lot of work, and it can be tiring to do for long periods. Because vigilance is so exhausting, meerkats must take turns keeping watch.

While one meerkat is on watch, the other meerkats can relax. They can forage or rest.[6] You may only see one meerkat on top of a rock or tree, but meerkats have a sophisticated system of vocal communication to "talk" with each other and coordinate the group's behavior.[8] The meerkat on watch is in communication with the rest of the meerkats in the burrow, making specific vocal calls to relay everything that she sees. If she spots a predator, she can make a unique alarm call to the rest of the group according to both the predator type and the level of urgency.[8] The rest of the group understands these vocalizations and prepares to either hide or attack.[7]

It makes sense that meerkats rotate through sentry duty when you consider how much work vigilance is. American physiologist Walter Cannon identified the sympathoadrenal system as being key to sensing changes in the environment or inside the body and preparing the body for a response, which became known as the stress response.[9] Studies have found that the sympathoadrenal system is involved in both vigilance and anxiety,[10, 11] telling the body how to respond to real or perceived threats. If you've ever felt anxious for more than a couple of days, you know how wiped out it can make you feel.

The body's stress response is physically exhausting, especially if it goes on for too long. Unlike adrenaline, which peaks in concentration and then quickly disappears, the glucocorticoids that are released by the adrenal cortex in a stress response can persist for hours after the threat has disappeared. They circulate throughout the bloodstream and cross back into the brain, where they activate a brain region called the locus coeruleus. This structure then sends a powerfully activating signal back to the amygdala, causing it to produce more corticotropin-releasing hormone, which leads to the secretion of even more glucocorticoids.[12] If the mind's perception of the stress is not changed, sometimes this positive feedback loop can go on and on, leaving the animal in a state of chronic stress.

You can see why always having to be on guard for threats would cause chronic stress in an individual, which is one of the reasons that soldiers of war often experience poor mental health. If vigilance goes on for too long, it can even change the size of certain brain structures and harm basic functions like reproduction. Cooperative vigilance, where the task of keeping watch is shared among multiple individuals, is therefore one of the main advantages of living in a group.

When a group of sailors makes a long ocean passage, they are completely reliant on one another for vigilance. They take turns on watch. At all times of day and night, one person is awake and at the helm, watching the horizon to make sure the boat doesn't crash. Even the others who are downstairs asleep are still semi-awake, though, rousing with any strange noise or unusual pattern. This routine of keeping watch and then barely sleeping, repeated over and over, gets to be exhausting. Often by the time the sailors arrive at a safe harbor where they can finally anchor, they are delusional, with their mental faculties down to their last thread. But the act of sailing together in this way can bond a group of people for life. When you're completely reliant on others for your survival, you become pretty close.

Like the sailors, meerkats rely on each other for their survival, and they live in tight-knit families. They even sleep on top of each other at night inside their burrows. Sleeping on top of each other causes the release of two key hormones involved with social bonding: oxytocin and vasopressin.[13] The frequent release of these hormones likely reinforces the group's cohesion, and oxytocin may even reduce stress levels within the group.[14] When cooperation is beneficial for the survival of a species, it often helps if social bonding is accompanied by an immediate reward, such as oxytocin-induced relaxation.

I glanced back at the sentinel meerkat on the tree. She looked aware but surprisingly relaxed. She was no longer standing tall but had sat down and was occasionally looking over at her fellow mob members on the ground. I heard soft murmurs amongst them, which sounded like short puppy whines. I couldn't understand what they were saying to each other, but I imagined they were working out who was up next on watch.

IV.

Meerkats are small-bodied animals compared with the eagles and hyenas trying to eat them, but by grouping together and taking turns on watch, they are able to effectively increase their level of protection against predators. Although we like to think of our early ancestors as dominating all the other animals of Africa, remains from 3.2 mya indicate that, like meerkats, we were also vulnerable to predation. Rather than the "Man the Hunter" scenario that is often painted in popular literature, it is more likely that we started out as "Man the Hunted."[15] And when you're being hunted, it helps to have someone else watching your back.

Enter Lucy (fossil specimen A.L. 288-1, *Australopithecus afarensis*), one of the most famous and complete skeletons we have from the prehistoric human lineage. Lucy was not human (*Homo sapiens*) but was a "hominin," a general name given by paleontologists to all species in the human lineage after our genetic split from chimpanzees and bonobos around 6 mya.[16, 17] From here on, you'll see the term hominin used to refer to our human-like ancestors. Lucy's body was discovered in 1974 in Hadar, Ethiopia. Paleoanthropologist Donald Johanson and his graduate student Maurice Taieb were listening to the radio while digging in the bed of a river when they discovered the remains of what looked like a small hominin. On the radio, "Lucy in the Sky with Diamonds" by the Beatles was playing, so they named her Lucy.[18]

Towering just over one meter tall and weighing a mere 28 kilograms,[19]

Lucy was not a natural-born predator. Originally she was thought to have been killed by a large carnivore, like a crocodile, but more recent analyses of Lucy's bones indicate that her death by predation is unlikely.[20] What appears to be more likely from the bone fracture patterns is that Lucy fell from a tree, extending her arms to brace herself before hitting the ground. Bone-into-bone compressive fractures are oriented along the long axes of Lucy's limbs at the shoulders, knees, and ankle joint, along with spiral fractures in the humeral and tibial shafts. These fractures suggest that Lucy fell from a considerable height and likely died upon impact or shortly after.[21]

Trees once abounded along the shallow, slow-moving stream where Lucy's remains were found.[20] But Lucy's features indicate that her kind were not tree dwellers. She had the long arm bones and powerful chest needed for tree climbing, but her broad pelvis, angled-in thigh bones, and compact feet were better suited for supporting all of her weight.[18] These features indicate that, unlike her great ape ancestors of the *Ardipithecus* genus, who only occasionally walked upright, Lucy spent large portions of her time walking on two feet.

If she was mostly a ground dweller, it begs the question of what Lucy was doing up in the tree. Perhaps she was hiding. When Lucy was alive, 3.2 mya, large predators roamed East Africa, including many megacarnivores who are now extinct, such as short-faced bears, giant hyenas, and several species of sabertoothed cats.[22] Strong indications from the fossil record lead us to the conclusion that hominins like Lucy were regularly hunted. As such, Lucy and her kind required social organization that made them inconspicuous as well as a system of coordinated vigilance.[23]

It's possible that because of their strong need for vigilance, Lucy and other early hominins developed a couple of adaptations that humans still carry today.

The first adaptation was searching for food in the middle of the day. At Lucy's stage of evolution, hominins had not yet developed the controlled use of fire or complex weapons. Therefore, to minimize their risk of predation, Lucy and her relatives might have chosen to go out in search of food during the hottest times of the day, when most predators are asleep or less active. If midday foraging was a survival strategy for our ancestors, then it could explain why humans have less fur and more sweat glands than any other primate. Most animals choose to carry out activities essential to their survival during the cooler parts of the day, but because they were on two feet and their predators were much larger, our ancestors likely didn't have that option. Instead, they probably took advantage of the hottest parts of the day, which minimized their risk of predation.[24]

Just because foraging was likely done in the middle of the day doesn't mean our ancestors spent the rest of their time sleeping. We see remnants of a second adaptation to predation in the sleep pattern of modern humans. Humans sleep the least of any primate, even though sleep is critical for our physical health and cognitive performance. Humans appear to be able to function with less sleep because we have a greater proportion of rapid eye movement (REM) sleep in our total sleep time than other primates do.[25] We have also adapted to living at latitudes with the greatest seasonal variation in light on Earth, and we have much greater flexibility in our sleep and wake times than other primates do, even in the absence of electric lights.[25] Our sleep pattern is short and flexible compared with other primates. For our understanding of cooperation, it's worth pondering why this is so.

One hypothesis to explain our weird sleep pattern is that early humans engaged in a type of cooperative vigilance throughout the day and night, like sailors who take turns on watch. The Social Sleep Hypothesis, proposed in 2021 by anthropologist David Samson, suggests that because early humans were small bodied and under constant risk of predation, they might have relied on sentinels, like meerkats do, to secure their sleep sites.[25] The use of sentinels would mean that at any given time, a few individuals would remain awake and on watch while the majority of the group slept. REM sleep, which modern humans do more of than any other primate, is the deepest phase of sleep. It's when most of our dreams occur. During REM sleep, the brain is extremely active, like an awake brain, but it is less sensitive to outside noises and stimuli. Also during REM sleep, the body is paralyzed, apart from some brief twitches of the muscles.[25,26] Noise blocking and muscle paralysis are not a good combination in the presence of predators. Had our ancestors like Lucy slept in synchrony, they would have all been vulnerable to predation at the same time. But if instead they took turns sleeping and keeping watch, they would have had protection during their most vulnerable states of sleep. If this hypothesis is correct, it could explain why humans have such flexible sleep schedules today. It may also explain why in some small-scale subsistence societies, like the Hadza in Tanzania, one or more adults will stay awake during the night while the others sleep.[25,27]

Sleeping on the ground is dangerous. Of all the primates, humans are the only ones who do so regularly, rather than sleeping in nests.[25] If Lucy's fall is anything to go by, then our *Australopithecus* ancestors likely made use of trees for sleeping when they could. Sleeping in trees might have provided them with highly mobile sleep sites to which they could withdraw at any time to get some deep, restorative sleep. But our ancestors like Lucy would have needed to move around a lot to get enough resources, so building permanent or even semi-permanent nests like those gorillas construct would have been impractical. Instead of building physical nests for sleeping, it could be that Lucy's social group provided the "nest." A sentinel sleep system within the group could have kept our ancestors safe while they were asleep, whether in trees or on the ground, even without a physical nest. It's plausible that the only way our ancestors could have been both mobile and safe during the periods of deep sleep their brains required was through cooperative vigilance.

Keeping an eye out for predators is hard work, and as meerkats have shown, it's a task best shared by an entire group. This may explain why our ancestors started working together in the first place. Once they established a sentinel sleep system, it could have created a positive feedback loop of emotions and behaviors that promoted in-group bonding.[25] If so, groups who were more cooperative likely had real fitness advantages, in terms of sleep quality, over less cooperative groups.

What gets interesting is how our cooperative instincts played out when the hominin population started to grow. In other words, does cooperation continue to work when small, intimate tribes grow into big, boisterous societies?

CHAPTER 4 REFERENCES

Further Reading

Sapolsky, R. M. (2004). *Why zebras don't get ulcers* (3rd ed.). Henry Holt and Co. https://us.macmillan.com/books/9780805073690/whyzebrasdontgetulcers

Sapolsky, R. M. (2017). *Behave: The biology of humans at our best and worst.* Penguin Press. https://www.penguin.com.au/books/behave-9780099575061

References

1. National Oceanic and Atmospheric Administration. (n.d.). *SARSAT Search and Rescue Satellite-Aided Tracking: Frequently asked questions.* Retrieved September 29, 2023, from https://www.sarsat.noaa.gov/faqs/

2. National Oceanic and Atmospheric Administration. (n.d.). *History of the SARSAT program.* Retrieved September 29, 2023, from https://www.sarsat.noaa.gov/history-of-the-sarsat-program/

3. Levesque, D. (2010). Cospas-Sarsat 1979-2009: A 30-year success story. *Cospas-Sarsat Information Bulletin Enclosure, 22.* https://www.cospas-sarsat.int/images/stories/SystemDocs/Current/bul22_final_enclosure.pdf

4. Fishman, J. A. (1982, October 24). Capsized trimaran changes thinking. *The New York Times.* https://www.nytimes.com/1982/10/24/sports/capsized-trimaran-changes-thinking.html

5. Clutton-Brock, T. H., Russell, A. F., Sharpe, L. L., & Jordan, N. R. (2005). 'False feeding' and aggression in meerkat societies. *Animal Behaviour, 69*(6), 1273–1284. https://doi.org/10.1016/j.anbehav.2004.10.006

6. Clutton-Brock, T. H., O'Riain, M. J., Brotherton, P. N. M., Gaynor, D., Kansky, R., Griffin, A. S., & Manser, M. (1999). Selfish sentinels in cooperative mammals. *Science, 284*(5420), 1640–1644. https://doi.org/10.1126/science.284.5420.1640

7. Graw, B., & Manser, M. B. (2007). The function of mobbing in cooperative meerkats. *Animal Behaviour, 74*(3), 507–517. https://doi.org/10.1016/j.anbehav.2006.11.021

8. Manser, M. B. (2001). The acoustic structure of suricates' alarm calls varies with predator type and the level of response urgency. *Proceedings of the Royal Society of London. Series B: Biological Sciences, 268*(1483), 2315–2324. https://doi.org/10.1098/rspb.2001.1773

9. Cannon, W. B. (1929). *Bodily changes in pain, hunger, fear, and rage: An account of recent researches into the function of emotional excitement* (2nd ed.). D. Appleton and Company. https://wellcomecollection.org/works/exuhm3xj

10. Maes, M., Meltzer, H. Y., Suy, E., Minner, B., Calabrese, J., & Cosyns, P. (1993). Sleep disorders and anxiety as symptom profiles of sympathoadrenal system hyperactivity in major depression. *Journal of Affective Disorders, 27*(3), 197–207. https://doi.org/10.1016/0165-0327(93)90007-7

11. Murison, R. (2016). The neurobiology of stress. In *Neuroscience of pain, stress, and emotion* (pp. 29–49). Elsevier. https://doi.org/10.1016/B978-0-12-800538-5.00002-9

12. Sapolsky, R. (2003). Taming stress. *Scientific American, 289*(3). http://www.jstor.org/stable/26060445

13. Lim, M. M., & Young, L. J. (2006). Neuropeptidergic regulation of affiliative behavior and social bonding in animals. *Hormones and Behavior, 50*(4), 506–517. https://doi.org/10.1016/j.yhbeh.2006.06.028

14. Uvnäs-Moberg, K. (1998). Oxytocin may mediate the benefits of positive social interaction and emotions. *Psychoneuroendocrinology, 23*(8), 819–835. https://doi.org/10.1016/S0306-4530(98)00056-0

15. Hart, D., & Sussman, R. W. (2018). *Man the hunted.* Routledge. https://doi.org/10.4324/9780429499081

16. Pontzer, H. (2012). Overview of hominin evolution. *Nature Education Knowledge, 3*(10), 8. https://www.nature.com/scitable/knowledge/library/overview-of-hominin-evolution-89010983/

17. Patterson, N., Richter, D. J., Gnerre, S., Lander, E. S., & Reich, D. (2006). Genetic evidence for complex speciation of humans and chimpanzees. *Nature, 441*(7097), 1103–1108. https://doi.org/10.1038/nature04789

18. Smithsonian Institution. (2022). *AL 288-1.* https://humanorigins.si.edu/evidence/human-fossils/fossils/al-288-1

19. Hendry, L. (n.d.). *Australopithecus afarensis, Lucy's species.* The Natural History Museum, London. Retrieved September 29, 2023, from https://www.nhm.ac.uk/discover/australopithecus-afarensis-lucy-species.html

20. Kappelman, J., Ketcham, R. A., Pearce, S., Todd, L., Akins, W., Colbert, M. W., Davis, C., Feseha, M., Maisano, J. A., & Witzel, A. (2019). Reply to: Charlier et al. 2018. Mudslide and/or animal attack are more plausible causes and circumstances of death for AL 288 ('Lucy'): A forensic anthropology analysis. *Medico-Legal Journal* 86(3) 139–142, 2018. *Medico-Legal Journal, 87*(3), 121–126. https://doi.org/10.1177/0025817219849367

21. Kappelman, J., Ketcham, R. A., Pearce, S., Todd, L., Akins, W., Colbert, M. W., Feseha, M., Maisano, J. A., & Witzel, A. (2016). Perimortem fractures in Lucy suggest mortality from fall out of tall tree. *Nature, 537*(7621), 503–507. https://doi.org/10.1038/nature19332

22. Treves, A., & Palmqvist, P. (2007). Reconstructing hominin interactions with mammalian carnivores (6.0–1.8 Ma). In S. L. Gursky & K. A. I. Nekaris (Eds.), *Primate anti-predator strategies* (pp. 355–381). Springer US. https://doi.org/10.1007/978-0-387-34810-0_17

23. Hart, D., & Sussman, R. W. (2011). The influence of predation on primate and early human evolution: Impetus for cooperation. In *Origins of altruism and cooperation* (pp. 19–40). Springer New York. https://doi.org/10.1007/978-1-4419-9520-9_3

24. Alex, B. (2019, January 18). Why humans lost their hair and became naked and sweaty. *Discover Magazine.* https://www.discovermagazine.com/planet-earth/why-humans-lost-their-hair-and-became-naked-and-sweaty

25. Samson, D. R. (2021). The human sleep paradox: The unexpected sleeping habits of *Homo sapiens*. *Annual Review of Anthropology, 50*(1), 259–274. https://doi.org/10.1146/annurev-anthro-010220-075523

26. Purves D, Augustine GJ, Fitzpatrick D, & et al. (Eds.). (2001). Stages of sleep. In *Neuroscience* (2nd ed.). Sinauer Associates. https://www.ncbi.nlm.nih.gov/books/NBK10996/

27. Samson, D. R., Crittenden, A. N., Mabulla, I. A., Mabulla, A. Z. P., & Nunn, C. L. (2017). Chronotype variation drives night-time sentinel-like behaviour in hunter–gatherers. *Proceedings of the Royal Society B: Biological Sciences, 284*(1858), 20170967. https://doi.org/10.1098/rspb.2017.0967

CHAPTER 5

Let's Work Together: Resilience

I.

Being out in nature on your own has got to be one of the best feelings in the world. What more do you need to be happy than the sensation of the wind pouring over your face?

I stood tall at the helm as my hair went wild in the wind. We were gaining speed in the Whitsunday Passage. I turned up the brightness on the chart plotter in front of my navel to see where we were headed—it looked like we were on course for the northwest tip of Hook Island, exactly where we needed to point. My grip on the helm was firm, and I had complete control.

I enjoyed standing behind the enormous helm. I discovered a wooden block on the deck next to my left foot, which appeared to be there to assist the skipper in standing upright when the boat was on a heel. Supporting my weight on it, I felt locked into position, like the universe had a perfect slot into which I fit.

"You wanna have a go?" I shouted to my husband. His eyebrows raised and he sprang up. I said, "I'm going to go downstairs," my voice still raised so he could hear me over the wind, " ... see what's happening with the zoos." He nodded, and we switched positions so he could take command of the helm.

I always know I can get my husband's support to sneak in a little work if it has to do with zoo animal welfare. He is the only person I know who is more of an animal lover than I am. Before leaving Brisbane, I was in the midst of a discussion with several zoos that had been forced to close to the public because of measures to contain COVID-19. These zoos were concerned about how they were going to keep feeding and caring for the animals, with no revenue coming in the door. Ever since they voiced their concerns, I had been replaying the worst-case scenario in my head, imagining majestic elephants and tigers and bears having to be put down.

I kicked off my shoes in the cockpit and spidered my way down the stairs, supporting myself with one hand on the deck and the other on a wooden handhold. Everything that we had neatly arranged on the galley bench—the box of Jatz crackers, the roll of paper towels, the 50 SPF sunscreen—had slid to one corner. I grabbed the bench and helped myself along and then collapsed into the cushioned seat. I pushed my feet against the folded saloon table to hold me in place.

I took my phone out of my shorts pocket and found the name of my colleague in my contacts. It rang a few times, and I worried I had caught him in a meeting.

"Hey, how are ya?" he exclaimed suddenly, as if I'd just returned from the dead. I smiled.

"I'm good. We're just doing a little sailing in the Whitsunday Passage. The weather is beautiful. How are things going there?"

"Pretty crazy, Jill. I'm not sure if you've seen the news." He paused, and when I didn't respond, he continued. "There was a heap of new coronavirus cases over the weekend in Sydney. South Australia has announced they're going to shut the border, and Queensland is talking about shutting too," he said.

"Shutting the border? How are they going to do that?" I asked. I couldn't fathom how a state within Australia could shut its borders.

"They're going to put police along the borders," he said. "Anyone coming in will have to do 14 days' quarantine in a hotel."

"That's crazy." I imagined being crammed in a hotel room for 2 weeks, breathing the stiff, musty air of a wall-mounted air conditioner and sleeping on cardboard pillows. "Why are there so many new cases in Sydney?" I asked.

"Apparently, everyone in Sydney decided to go out and party, even though they were told to stay home. You should have seen the footage—Bondi Beach was completely packed. People all over the beach, right on top of each other, no one wearing masks. Everyone's gone bonkers."

"Wow," I said, chuckling. I pictured the beach scene from a bird's-eye view with amusement. Glistening bodies right next to each other on Bondi's golden sands, viral particles passing between them. I thought, *It's just like Australians to go for one last hurrah on the beach before the government forces them inside.*

"What about the zoos ... what's going on with them?" I asked.

"They're all closed now. They've got no money coming through the door but huge ongoing costs for the animals. We're having our first meeting of the COVID-19 animal welfare response group tomorrow," he said. "The governments are still introducing regulations, and people are starting to worry that they're not going to be able to move animals across the state borders. We'll see what happens with the zoos ... they're compiling a list of food they need for the animals. Maybe some of the other industries could donate carcasses."

"That's amazing," I said. Inside I felt a twinge of regret that I wouldn't be there for the first meeting. Just as all the people looking after animals in Australia were coming together to support each other through the crisis, here I was flitting about the Whitsundays. My stomach felt hollow.

I guess on some level, it made sense that the people at Bondi Beach were coming together in a time of crisis. When animals feel unsafe in their environment, they often huddle in closer for safety. *Best we stick together*, is the mindset. *Less danger to any one of us.* Maybe sticking together is not why people were flooding to Bondi Beach, but it was funny to think that even as cognitively advanced as we supposedly are, even knowing there was a deadly virus going around, our instinct when faced with the unknown was to herd.

"I bet you're glad you didn't go to Mexico," my colleague said. "You'd probably be stuck there for who knows how long."

"Well," I said, imagining the charred corn tortillas holding an assortment of juicy meats, "there are worse places to be stuck." He laughed. My mind returned to the Bondi Beach scene, packed with people. Although I was happy to be out on our little boat, safe from coronavirus, I got an eerie feeling that the world back on shore was caving in. *How will we feel*, I wondered, *if society is in shambles when we return?*

II.

If there's one place I would flock to in a pandemic, it's not Bondi Beach. It's Mexico. Mexicans are not prone to panic. Their society has been around for a long time (some 30,000 years, by latest estimates), and it seems they are used to absorbing all kinds of shocks.

The first time my husband and I were in Mexico, there was an earthquake. We were staying in the seaside town of Puerto Escondido in the state of Oaxaca. My husband wanted to surf on Mexico's Pacific coast, so he rented a Mini Mal from a shop at the nearby Playa Zicatela. With his board under his arm, he walked next to me along the sandy main road towards the beach. As we entered the beach, I noticed some local boys laughing and pointing at a thin, wooden pole in the sand. I didn't understand what was happening, but they pointed downwards to the sand, which was wobbling like a plate of Jello. The ground beneath us was moving.

The tremor seemed mild and soon passed, so my husband proceeded down the beach to our planned activities. He took to the waves and I to a lounge chair, courtesy of a quiet palapa eatery on the sand. In every direction, the beach was devoid of other tourists. In front of me, a young man holding a fishing net waded in the waves. After several minutes, he came ashore with his catch: half a dozen fish. He carried the net through the restaurant and back to the kitchen. Then he returned to take my order. He had kind eyes and a familial manner; I got the impression he was the oldest son of the owners. I asked him for a mojito, even though the weather was dreary, and it probably wouldn't be the most enjoyable mojito. He returned to the shelter to start work on it. I squinted to make out my husband's black hair bobbing among the gray waves. The cloud cover and the mist made it chilly, and I put my beach towel over my thighs to stay warm. I thought, *At least I won't get sunburned.*

Not one minute after the young man brought me an icy mojito, it started to sprinkle. I grabbed my towel and glass and headed back to the straw-roofed shelter, where it was dark and desolate. Aside from the young fisherman/waiter, there was just one other person, a plump woman in the kitchen listening to talk radio as she heated the fryer.

Carrying his phone in one hand, the waiter moseyed over to me. He asked in Spanish,

"Did you see the tremor?"

"Yes," I said. "Do you know where it was?" I asked.

"In Puebla," he said.

"Was it strong?" I asked.

"Yes, very strong," he responded. "Look." He put his phone in front of me and pressed Play. The video showed the crowded skyline of Mexico City, the second largest city in the Americas. And in an instant, the skyline began to disappear. Building after building collapsed. It looked just like 9/11 when the twin towers suddenly fell in on themselves—*poof*—but it was happening all over the city. In one moment, these towering icons of human progress were there, and the next moment, they were gone.

I had never seen anything like it. I pulled out my phone and searched for Puebla on Google Maps. It was 11 hours away by car, close to Mexico City. I wondered whether it would cause a tsunami—*how does that even happen?* My thoughts raced to my husband out there bobbing among the waves. I Googled "tsunami Mexico," but there was no mention of it. Then, just as nonchalantly as the waiter had raised the topic of the earthquake, he put his phone away, grabbed his fishing net, and headed back down to the water's edge. I thought, *If he's not worried about a tsunami, maybe I shouldn't be either.*

When my husband came in from his surf, I told him about the Puebla earthquake. I watched his eyes widen before narrowing into survivor mode. The country would be in a state of disaster. Would we be safe, would buildings collapse, would we be able to get out?

Our first mission was to return the surfboard. As we walked down the sandy main street, the place was just as sleepy as before. People were still chatting in relaxed tones, going about their business at a snail's pace. It was so bizarre. But then, this was Mexico. Did I really expect any different?

If the waiter hadn't shown me that video, I never would have guessed that an earthquake of magnitude 7.1 had just occurred. I remembered hearing that Mexico City was prone to earthquakes because it sits on the bed of a historic lake. In high-school Spanish, we had learned that when the ancient Aztecs asked their gods for a sign of where to build their holy city, Tenochtitlan, the sign of an eagle perched on a cactus and holding a snake is said to have appeared in the middle of a lake. They built their city on the lake by interweaving reeds with stakes to create underwater fences and filling these fences with soil. Not a very solid foundation for Mexico's modern capital, but as Mexicans say, *así* es la vida—such is life.

The Puebla earthquake that we witnessed happened, coincidentally, on the 32nd anniversary of the horrific 1985 Mexico City earthquake. The 1985 earthquake was actually two earthquakes. There was so much pent-up energy in the interaction of the tectonic plates that it was released in two hits. The first quake was at 7:18 a.m. on September 19, registering 8.1 on the Richter scale. The second one hit at 7:38 p.m. the next day, registering 7.5. At least 5,000 people were killed in the 1985

Mexico City earthquake, and at least 14,000 were injured. Over 100,000 building units were damaged, leaving 2 million people homeless. Billions of dollars' worth of material damage was done, including to important federal government buildings, financial centers, and key communication centers.[1] It was a disaster by any definition. And it was the perfect opportunity to understand how societies cope with chaos.

Sociologists at the University of Delaware's Disaster Research Center watched the 1985 Mexico City earthquake from afar, through television and newspapers. The Center's leaders, Russel Dynes and Dennis Wenger, understood natural disasters well. They knew how to plan for them and how to manage them once they happened. But not being in Mexico, they assumed there was little they could do.

Just two days after the 1985 earthquake, Dynes and Wenger received a visitor. His name was Alejandro Garnica Andrade. He had been sent by the Instituto de la Comunicación, a Mexican survey research center, because they urgently needed some expert advice on how societies respond to disasters. Most of the physical damage from the earthquake was incurred in three neighborhoods of Mexico City, but the societal disruption was widespread. Dynes and Wenger spent the next 3 days with Andrade, discussing the societal aftermath of natural disasters. Equipped with all this knowledge, Andrade returned to Mexico City and conducted a survey on the response of Mexico City's population to the earthquake. At just 3 weeks out from the disaster, it was the earliest survey of a large population ever done following a major disaster.[1]

When a natural disaster happens, Dynes and Wenger had explained, it is common for the affected population to grow critical of the government soon after the event. They will often complain about the government's being too slow to respond or poorly executing the response. But

Andrade found this was not the case in Mexico City following the 1985 earthquake. Yes, there were problems, admitted the survey respondents. Around 38% said that the problem of homelessness was magnified by the disaster. Another 13% said that the disruption of water and other services in some neighborhoods was a problem. But despite these problems, there was little criticism of the government's response (only 6% of respondents had a negative view). Their favorable assessment of the situation extended to other groups as well, with most residents reporting that they were happy with the actions of the local government, the army, the president of Mexico, the volunteer groups, and even the police.[1]

Dynes and Wenger were startled by the Instituto's findings. It was a societal reaction unlike anything they had ever seen following natural disasters in other parts of the world. So when the U.S. National Science Foundation made funding available to study this disaster, Dynes and Wenger asked the Mexican sociologists at the Instituto whether they would collaborate to re-analyze the survey results and conduct a follow-up survey one year later to see if the Mexicans' attitudes about the disaster had changed.[1]

The Instituto agreed to the study, and Andrade flew to Delaware for a week to plan the research and design the survey. After that, four members of the Disaster Research Center flew to Mexico City to train the interviewers. On the first anniversary of the 1985 earthquake, they carried out the second survey. The Mexican sociologists then compiled and coded the data and mailed computer tapes to the Disaster Research Center for the American sociologists to analyze.[1]

What the second survey, conducted one year out from the earthquake, revealed about Mexican society was even more extraordinary than the first.

A year on, and the Mexico City residents were still not criticizing the government's response or complaining about disruptions to utilities or services. Many went without running water for a considerable period, and yet the majority of survey respondents said that the restoration of water service was handled well. Regarding interruptions to the telephone service, survey respondents even praised the telephone company for being well organized and showing compassion during the disaster.[1]

It would seem that, instead of spending their time criticizing the post-disaster efforts, Mexico City residents got to work. The survey revealed that one in every eight adult residents volunteered during the post-earthquake period. Compared with other natural disasters around the world, this was a massive outpouring of volunteerism by local residents.[1]

Even more interesting was how the Mexico City residents pulled together after the disaster. Of the survey respondents, 10% left their own homes to stay with family and friends. This is not unusual for post-disaster situations around the world. What was surprising, though, is that among the Mexico City residents who moved, only around 15% said that they moved because of severe damage to their homes. The vast majority of people who moved in with family and friends following the earthquake did so because of psychological or economic reasons, not because they physically couldn't stay in their own homes. Of the people who moved in with family and friends, the majority (60%) stayed for more than 2 weeks and up to a month.[1]

Previous studies of natural disasters have found a high degree of community solidarity immediately following a natural disaster, but this solidarity tends to dissipate over time. The 1986 Mexico City survey found that this was not the case in Mexico. Three out of every five survey respondents reported that the earthquake actually increased the trust that they had in other citizens.[1] It would seem that instead of causing Mexicans to blame authorities or each other, the earthquake actually brought Mexicans closer together.

What happened in Mexico after the 1985 earthquake and what we saw during the 2017 earthquake says a lot about Mexico as a society. In response to major disruptions in their environment, Mexicans do not turn on each other—they come together. Archival material reveals that, even in colonial times, periods of crisis in Mexico did not result in societal breakdown.[2] Instead, just like with the 1985 earthquake, the shared experiences of the crisis created a greater sense of cooperation.

To this day, Mexican culture perpetuates this kind of "cooperative coping." When they need support, Mexicans go to their families and extensive networks of friends and neighbors. They tend to have high levels of membership in community associations, large social networks, and a general belief in reciprocity. These features of social life contribute to Mexicans' generosity towards each other,[3] and together they form what the political scientist Robert Putnam would call "social capital."[4] Mexicans may face many challenges in terms of economic capital, but as a people, they are rich in social capital. Their social capital makes their society more resilient when it comes to facing crises, like earthquakes. It should come as no surprise that Mexicans' social capital has helped them to cope with uncertainty during more recent crises, such as the COVID-19 pandemic.[5]

Looking at the way in which Mexicans cooperate in the face of crisis, maybe now we can understand how our hominin ancestors might have first formed their own resilient societies based on cooperation.

III.

There is a big difference in scale between the kind of cooperation that happened within the small groups of our hominin ancestors and the kind of society-wide cooperation that happened in Mexico City after the 1985 earthquake. So how were our ancestors able to progress from cooperating on a family scale to a societal scale? And why? What benefits did widespread societal cooperation bring them? By now you should not be surprised that I turn to other animals for the answers.

The first time I saw sea lions in the wild was actually in Mexico. As our dive boat approached the parched island of Espíritu Santo off the Baja California peninsula, I saw what looked like shiny chocolate sausages draped over the rugged pink and orange rocks of the shoreline. The water lapping against the cliffs was neon blue, like the blue freeze pops we used to slurp during summer as kids. Also reminiscent of my summers as a kid at the Kansas City Zoo was the distinctive smell of California sea lions (*Zalophus californianus*).[6]

I donned my snorkel, mask, and fins and rolled off the side of the boat into the crystal water. Less than a minute later, the first one approached. I saw a big splash of bubbles on the surface at the rocks' edge. He dove like a fighter jet, his fins extended to each side. Once he was a few meters under, he started to swim in front of me, spinning belly up and bending his blubbery body in graceful arcs and figure eights. He came closer with his pronounced, infant-like eyes, bringing his face close enough for me to admire his whiskers. I was startled, thinking he might crash into me. Just before he did, he darted to one side, looping back to where he had started.

Another splash broke the surface. A second blubbery body came into view. He began to play with the first one, rolling around provocatively and tagging him before swimming away. The first one engaged him without hesitation. The two spun around each other, back-to-back, their mouths gaping open. It appeared to be something they had done many times before, like a choreographed dance. They circled in front of me, approaching every few seconds. Perhaps they hoped I would join their dance, but I was too slow to respond. Mesmerized, I was busy trying to capture and lock in my memory every millisecond of the experience.

This type of play is common in sea lions, I later learned. Not all animals play, or at least not that we know of, but sea lions are a playful bunch. They have even been seen playing with whales. The social play that I witnessed among the sea lions at Espíritu Santo took me by surprise. I found it bizarre how casually they sought to involve me, a strange-looking human. After my snorkel with them, I became really intrigued by animal play. In good Darwinian style, I wondered, *How did these animals evolve to behave in this way?* Surely it couldn't be a good survival strategy, approaching total strangers and asking them to play, right?

When I looked deeper into the context of sea lions' evolution, I realized that perhaps their gregariousness wasn't as crazy as I'd first believed.

For a vivid illustration, we turn to the Galápagos Islands. Sitting less than one degree south of the equator, the Galápagos Islands are not somewhere you would normally expect to see a large-bodied marine mammal. Sea lions use a lot of energy. Where there is an abundance of fatty fish, there is an abundance of sea lions, so the colder waters of northern California, British Columbia, and Alaska sustain large populations of sea lions.[7] But as Darwin noted in the 1800s, all kinds of unexpected animals have ended up in the Galápagos.

The Galápagos sea lion (*Zalophus wollebaeki*) is endemic to the Galápagos Archipelago, but the population size there is only 14,000–16,000 individuals.[8] Apart from the latitude, the population is small in this region because of exposure to the El Niño Southern Oscillation, which is a periodic cycle of warm currents that pass through the tropical Pacific Ocean.[9]

El Niño events have been occurring for thousands of years. When El Niño comes, roughly every 2–7 years, sea surface temperatures increase by 2–9 °C along the tropical Pacific coast of South America. The warm waters stick around for anywhere between 9 months and 5 years.[10, 11] During El Niño, fishermen off the coast of Peru and Ecuador have seen fish populations collapse. This is because the warm surface currents of El Niño reduce the upwelling of cold, nutrient-rich water from the ocean floor, so fewer phytoplankton can survive. In such conditions, the animals who eat the plankton, like clams, are less abundant, and the fish who eat these animals are deprived of food.[10] The situation of nutritional stress created by El Niño results in massive declines in marine life, which is what caused the world's largest fishery to collapse in 1972 (the Peruvian anchovy).[12] Due to repeated impacts on fish supply, El Niño events have caused the Galápagos sea lion population to shrink by half in the last few decades.[9]

In 2015, sea lion biologist Diego Páez-Rosas was working at the Galápagos Science Center on the Island of San Cristóbal when he got a call. It was Franklin Arreaga from the COPESPROMAR Fishing Cooperative. Arreaga told Páez-Rosas that he had received some information from local fishermen about an unusual hunting behavior they had seen in the male sea lions inhabiting the northern region of Isabela Island.

The behavior reported by the fishermen was different from the typical hunting strategies of male sea lions, who were known to be territorial and to hunt alone.[13]

Intrigued by the fishermen's account, Páez-Rosas hopped on a motorboat with a few of his team members and puttered to the top side of Isabela Island. Páez-Rosas eased the throttle as they approached a cluster of rocks called Piedras Blancas. Páez-Rosas could see a small group of sea lions resting on the rocks exactly where the fishermen had described.[13]

The team pulled ashore near the group but not too close, and they began to walk along the coastline, observing the sea lions from different vantage points. They identified three adult males and three adolescent males, who all appeared to be in good health. The males spent long periods hauled out on the rocks, resting, and every so often they would plunge into the water to hunt.[13]

It was difficult to see what the males were doing when they were hunting in the water, so the next time the researchers visited, they brought a drone to observe the sea lions from 15 to 20 m above the water surface. The drone footage revealed that the sea lion males were on the hunt for yellowfin tuna. Just as the fishermen had said, they were carrying out an unusual behavior to get the tuna. These meaty pelagic fish, at least one meter in length, would have been a jackpot for a hungry sea lion in the Galápagos. But yellowfin tuna are quick, swimming at 64 km/h, which makes them almost impossible for a single sea lion to catch. On their drone footage, Páez-Rosas and his team spotted the male Galápagos sea lions using an innovative strategy that would overcome this hurdle.[13]

Two or three of the adolescent male sea lions would leave the bay and move offshore together. The team would locate a school of yellowfin tuna and begin zigzagging through it so as to disturb the school of fish. Once a few tuna had been separated from the school, the adolescent male sea lions all increased their speed, driving the fish into the shallow bay, where three of the adult male sea lions were standing by. As the tuna were corralled into the bay, they would try to escape, but each time they tried, one or two of the adult male sea lions would come from behind and block their exit. Once the yellowfin tuna were trapped, the adolescent male sea lions herded the fish towards the shore. In the shallows, the trapped tuna would jump wildly, their iridescent bodies emerging from the water, and the adult male sea lions could easily chomp them in mid-air. The adults would feed, leaving a large part of the tuna for the adolescents. Using this cooperative hunting strategy, the male sea lions were able to capture big fish that would have been impossible to capture alone.[13]

In most sea lion colonies around the world, male sea lions hunt alone and are protective of their hunting grounds. This is what makes the case of the sea lions in the Galápagos so odd. Although cooperative hunting is not the norm for sea lions, Páez-Rosas and his team discovered that when the environment demands it, male sea lions will coordinate their actions to catch prey.[13]

Where the availability of high-quality food is unpredictable, like yellowfin tuna in the Galápagos Islands, animals often need new strategies to improve their chances of eating.[13] Over time, repeated phases of nutritional stress, such as those caused by El Niño events, can lead to a population of animals developing novel strategies to obtain food.[9]

For cooperative hunting to work, the Galápagos sea lions need to be able to coordinate their actions. They need to be able to work towards a common goal and adjust their individual actions to align with the actions of others—no simple behaviors. How did the Galápagos sea lions develop such sophisticated abilities?

In the 1970s, primatologist Frank Poirier put forth one theory. He suggested that animals likely develop sophisticated behavioral patterns, like those required for cooperative hunting, by playing together when they are young. Through play, Poirier theorized, animals combine their existing behavioral patterns into new combinations, like twisting a Rubik's Cube.[14] Ecologist Roger Gentry has observed that from the time sea lion pups are 8 weeks old, they begin to combine their play behaviors into sequences or patterns. Although basic postures, gestures, and sounds are hardwired in them, sea lions learn the combination of these behaviors through play.[15] If Poirier and Gentry are correct, then the Galápagos sea lions' cooperative hunting strategy was likely the result of behavioral combinations that they had developed as playful pups.

As sea lion pups grow, they prune and hone their behavioral combinations to suit their environment. Movement patterns that work best in any given situation are retained, while less effective patterns are lost.[15] The result is a large repertoire of behavioral combinations that they can draw on later in life. As adults, sea lions piece together these sophisticated patterns of behavior to adapt to changes in their environment. According to Poirier's theory, play gives these animals behavioral flexibility, preparing them to adapt to a wide range of circumstances and environments.

Play is not only important for sea lions' flexibility as individuals, but for the flexibility of their groups. Through play, Galápagos sea lion pups learn how to coordinate their actions with each other, how to communicate most effectively with each other, and how to express aggression appropriately without being rejected from the group.[16] It's possible that the sea lions I saw in Mexico were doing this sort of testing of others' reactions by engaging me in their play. Playful behavior by sea lion pups may lay the foundation for refined social behavior in adulthood, including the social perception needed by Galápagos sea lions to hunt successfully together in an unpredictable environment. In other words, play begets the capacity for cooperation, and cooperation begets resilience.

These cooperative foundations of resilience, which Galápagos sea lions and many other animals develop early in life, would have also been useful to our early ancestors in their highly unpredictable environments. Through an iterative process, like play, animals can combine and refine simple social behaviors to allow for more complex social behaviors to emerge. It's likely that our ancestors refined their complex social behaviors to form their own bustling societies—societies that could weather any number of crises.

IV.

The sea lions in the Galápagos clearly had some innate social tendencies, but pressures in their environment squeezed them into a situation in which they depended on those social tendencies just to survive. Is it possible that our ancestors experienced a similar environmental "squeeze"?

Several lines of evidence indicate that yes, hominins experienced an intense environmental "squeeze" around 2.8 mya, the result of which was the origins of human society as we know it today.

The first change was a shift in the timing of the ice ages. The ice ages used to come and go largely in step with the wobble of Earth's axis.[17] Our planet isn't at all like a plastic globe, which spins perfectly on a point. The real Earth wobbles as it spins, a phenomenon known as axial precession. This means that the direction of Earth's axis relative to the fixed positions of stars moves in an elliptical cycle of about 23,000 years.[18]

If the Earth didn't wobble, then the north and south poles would take equal turns in building up sea ice during their respective winters, with the ice melting in summer. But because the Earth does wobble, it allows ice to grow faster at one pole than at the other. This growth of sea ice results in greater reflection of sunlight, which makes the Earth colder and colder and plunges the Earth into a glacial period, known as an ice age.[19]

Prior to 2.8 mya, the ice ages were coming and going every 19,000–23,000 years, corresponding with the Earth's wobble. But around 2.8 mya, something strange happened. The pattern of the ice ages suddenly slowed to once in every 41,000 years. Scientists believe this change happened when the Earth's tilt (obliquity) took over as the dominant force driving ice ages in lieu of the Earth's wobble.[20]

Ice ages have a major effect on our global climate. During an ice age, the whole planet is cooler and drier than it would be normally, with much of the moisture being tied up in ice. Between ice ages, this reverses, with warmer temperatures and lots of rainfall.[19] So around 2.8 mya, when the timing of the ice ages changed to once in every 41,000 years, it was followed by a window of increased variability. The African landscape became a pendulum, swinging between periods of prolonged drought and great floods.[20] During this erratic transition, many forest-dwelling mammals in East Africa underwent extinction. Grazing animals who could digest dry grasses, such as wildebeest, gazelles, antelopes, and giant warthogs, gradually became more abundant.[21] Our hominin ancestors, too, had to adapt to survive this tumultuous remodeling of Africa.

Prior to 2.7 mya, almost all upright-walking hominins were like Lucy, members of the *Australopithecus* genus. But in the few million years following the transition in ice age timing, two new lineages of hominins emerged: 1) a robust sub-genus of *Australopithecus*, called *Paranthropus*, and 2) the earliest members of our own genus, *Homo*.[20]

Our genus was stumbled upon by archaeologists Louis and Mary Leakey, when they found two teeth at Olduvai Gorge, Tanzania, in 1959. Later they found parts of a boy's skeleton at the site and fossils of other individuals.[22] As the Leakeys examined the fossils, they saw a creature who looked more human than *Paranthropus* or, in fact, any other ape who had come before. Five years after their discovery of the first remains, the Leakeys and colleagues attested to the scientific community that the unearthed individuals did not belong to the *Australopithecus* genus. Due to the shape of their molar teeth and large cranial capacity relative to body size, they belonged to the *Homo* genus—the same genus as modern human beings.[23] Because the Leakeys had found the *Homo* remains next to thousands of stone tools at Olduvai Gorge, they named the species *Homo habilis*, meaning "handy man."[24]

The thousands of tools surrounding *Homo habilis* were a more advanced variety of flaked stone tool than what had come before. More primitive stone tools used for battering and smashing had existed as long as 3.3 mya, but sharp-edged stone tools only emerged around

2.6 mya.[25] Archaeologists call the method used to make these sharp-edged tools Oldowan technology—it involves using hammerstones to strike large stone cores and produce sharp flakes.[26] Tools made by Oldowan technology include choppers and scrapers, which were most likely used for cutting, pounding, and crushing tough plants and animal tissues.[17] Although tools had been used by hominins before, and certainly many primates use tools today, *Homo habilis* are considered the first to have manufactured complex stone tools, which requires the use of tools to make other tools.[25, 27]

Putting all these events together, what picture of hominin society can we see? What I see through my interpretative lens is that some kind of environmental "squeeze" took place in East Africa around 2.8 mya. Our ancestors might have experienced nutritional stress following the transition in the timing of the ice ages, akin to what Galápagos sea lions experience during El Niño events. Our ancestors likely lacked access to their normal forest foods of fruits, seeds, and small animals and faced a smaller range of nutritious options.[28] And like the Galápagos sea lions, our ancestors needed to trial new strategies of acquiring food if they were to survive.

At this time, two different nutritional strategies were taken by hominins. Studies of jaw morphology and teeth wear indicate that the robust lineage of *Paranthropus* hominins coped with the nutritional stress by taking a specialist's approach, whereas hominins of the *Homo* lineage engaged in a more flexible approach. *Paranthropus* became better suited to the tough grasses and roots that were proliferating throughout Africa. *Paranthropus* evolved to have strong, protruding jaws and large molar teeth, presumably for chewing large quantities of low-quality, fibrous grasses, as well as nuts and hard-shelled seeds.[29] In other words, as the environment became harsher, *Paranthropus* became tougher.

Homo habilis, on the other hand, became less tough but more versatile. Molar shape and dental microwear studies indicate that *Homo habilis* generally ate softer foods than *Paranthropus*.[30] While *Paranthropus* had large jaws and teeth, *Homo habilis* had flatter faces with smaller molars and larger front teeth, like modern chimpanzees and orangutans.[30, 31]

But with disappearing forests, where were *Homo habilis* finding these soft foods? There is some evidence that what *Homo habilis* lacked in brute strength, they made up for with their use of tools. Stone artifacts and cutmarks on bones found at Olduvai Gorge have led some paleontologists to believe that *Homo habilis* were using sharp-edged tools to process large animal carcasses.[26] Although some of the marks on the bones are toothmarks of other animals, analysis by scanning electron microscopy indicates that over half of the cutmarks are the result of sharp tools used by hominins.[32]

The bones that *Homo habilis* were processing could have been obtained by scavenging leftovers of carcasses on the savanna, which wouldn't have required much cooperation at all. But had our ancestors only scavenged, it would have been hard for them to get the number of bones they needed to survive. Organs obtained from fresh kills could have been a nutritious supplement to bone marrow, but finding carcasses with organs still in them would have been difficult. Most African predators consume the abdominal organs first, then move on to muscle meat, leaving only the skull and bones behind. Plants like roots, beans, nuts, and tubers would have been available at different times of the year. These plants are rich in nutrients, but without cooking, the bioavailability of nutrients in them is low, and there is no evidence of controlled fire use 2.8 mya.[33] So *Homo habilis* must have depended on animal carcasses with organs or some meat still on them, but for a lanky 100-centimeter animal who lacks the physical prowess of a predator, the question is how.[24]

In one of his later books, *Descent of Man and Selection in Relation to Sex* (1871), Darwin proposed that rather than merely scavenging, perhaps our ancestors adapted to their new environments by coordinating their actions to hunt. Most paleontologists agree that from around 2.8 mya, our hominin ancestors were somehow acquiring large animal carcasses and using stone tools to extract nutritious meat, ligaments, and bone marrow.[33] Muscle meat from carcasses contains 2–5 times more fat than organs do, so it would have been a rich source of bioavailable protein, iron, and vitamins for our ancestors. Muscle meat also concentrates and stores vital nutrients, so while many of the plants on which our ancestors depended died out in times of drought, the animals of East Africa likely became an important store of nutrients to sustain our ancestors' populations.[33]

Scientists still debate whether our ancestors obtained large animal carcasses through active hunting, confrontation with other predators to steal their recent kills, or passive scavenging of abandoned carcasses. Of these strategies, active hunting and confrontational scavenging are the only ways of obtaining large amounts of muscle meat. Both active hunting and confrontational scavenging require cooperation. By venturing out in small groups, which might have included distant relatives or non-relatives, perhaps our ancestors hunted animals or confronted predators to capture large carcasses with meat still on the bone.

In both the active hunting and confrontational scavenging scenarios, our ancestors would have had some heavy, meaty carcasses on their hands. Artifacts and bones found at Olduvai Gorge indicate that the average *Homo habilis* individual weighed a mere 35 kg, whereas the most commonly acquired carcasses weighed 115–340 kg.[28] Some of the carcasses found at Olduvai Gorge were the kills of other carnivores, but the majority are thought to have been transported and deposited there by hominins.

Dismembering a whole carcass in an open field would have attracted several other predators who were much larger than *Homo habilis*. In fact, *Homo habilis* was smaller than eight of the 12 large carnivores inhabiting the African savanna at the time.[28] So, as many carnivores do when hunting sites are competitive, our ancestors would have prioritized getting carcasses to safety as quickly as possible.[28] Hominin fossils, animal carcasses, and stone tools have all been found in concentrated spots, indicating that our ancestors had preferred sites for processing and eating.[33] The diversity of animal fossils found at Olduvai Gorge indicates that our ancestors transported animal carcasses for several kilometers back to their preferred sites.

Transporting a carcass for several kilometers is no small task. It would have required several people to carry the carcass together, possibly while running.[28] Carrying something heavy while running requires a high level of joint focus and coordination.[33] Even something as simple as carrying a heavy box with another person requires coordinating action and social perception.[34] For cooperative carcass transport to have worked, our ancestors would have needed to have a common goal in terms of the direction they were headed, predict what other individuals in the group were about to do, and adjust their own behavior accordingly.[16] These cooperative skills must have developed throughout their lifetimes, starting with play as children. Those groups who were better at cooperating would have had a better chance of transporting large carcasses to safe spots where they could dismember them. Thus, they would have had a better chance of survival.

The picture that emerges from these prehistoric lines of evidence is that our ancestors developed an intricate system of acquiring food, a system based on cooperation. If our ancestors were working in large groups to obtain and process food around 2.8 mya, then it would have constituted the earliest form of hominin society—a cooperative group that extends beyond one's own family.

Like humans, most other primates invest significant portions of their day in maintaining social relationships. One of the ways in which non-human primates like chimpanzees do this is through social grooming, where one chimp will sit next to another and carefully remove debris from the other's skin and fur. Although this behavior has a hygienic function, evolutionary psychologist Robin Dunbar has found that Old World monkeys and apes spend significantly more time grooming each other than what is required for hygiene alone.[35] His theory, which is an extension of primatologist Cor Goosen's theory,[36] is that primates groom each other mostly so that they can stay in each other's company for a long period. Dunbar has found that, at least in Old World monkeys and apes, the larger the group size, the more time individuals spend grooming each other each day. This suggests that social grooming plays a role in group cohesion.[35]

Group cohesion is fairly straightforward when groups are small—you scratch my back, and I'll scratch yours. But as hominin groups grew larger, which is thought to have happened around the emergence of the *Homo* genus,[37] it would have been impractical for them to spend the time needed to groom everyone individually in order to maintain good relationships. In group sizes larger than 40, our ancestors would have had to compromise on their social grooming time in order to continue other important activities for their survival, like feeding and travel.[38] Dunbar suggests that in these larger groups, cohesion had to be maintained by other means. So how, in the face of enormous ecological pressures, did our ancestors manage to keep their groups together?

One possibility is that they got funnier.

Laughter is thought to have evolved as a way of meeting the increased demands of social bonding.[37] According to Dunbar, we first started laughing some 2.5 million years ago, in the early days of the *Homo* genus. Around this time, our ancestors' expansion into new territories required them to have groups of more than 40 individuals in order to stay safe within predator-risky habitats. Soon their groups became too large for classical primate forms of social grooming to be time efficient. Laughter—a form of group chorusing—therefore took the place of sifting through one another's fur.[37] Laughter is a similar behavior to the play invitation pant of Old World monkeys and does not require the motor ability for speech, which likely came later. Like social touch, laughter triggers the release of endorphins in the brain of everyone who is laughing, which promotes a sense of belonging and feelings of closeness with others. But laughter is nearly 3 times as effective as touch for stimulating the endorphins that promote group bonding. For hominins, this would have made laughter a better tactic for building cohesion within large groups.[37]

Another way to look at laughter is as a form of social play. The functions of play for animals are still poorly understood, but one of the proposed functions is to discharge stress through physical exertion. Whether play fighting like the Galápagos sea lions or laughing heartily in a gathering like our ancestors would have done, it could be that group play provides social animals with a periodic release of stress. This regular offloading of stress may be necessary for animals to thrive in sparse or unstable environments.

Hominin societies based on cooperation were likely "squeezed" out of unpredictable environments. Just like the Galápagos sea lions, our *Homo habilis* ancestors needed to twist their behavioral Rubik's cubes in just the right combinations so that they could harmonize their actions. Forms of group play, such as laughter, would have facilitated their harmony.

Our ancestors needed to cooperate well enough to obtain, transport, and process meaty carcasses to feed their growing tribes. And at the end of the day, they needed to share a good laugh just to stay sane. Their ability to do so wasn't superfluous—successful cooperation meant resilience for their societies. Resilience would mean life over death in a frequently changing climate. It is because of their laughter, their cooperation, and their resilience that we are alive today.

Each animal in Africa had its own way of dealing with the environmental chaos of 2.8 mya. Cooperation was ours. In fact, cooperation would make hominins so resilient that soon we would no longer be tethered to one environment or even to one continent.

CHAPTER 5 REFERENCES

Free Resources

More information about Milankovitch (orbital) cycles and how they affect Earth's climate can be found at NASA's Global Climate Change website: https://climate.nasa.gov/news/2948/milankovitch-orbital-cycles-and-their-role-in-earths-climate/

References

1. Dynes, R. R., Quarantelli, E. L., & Wenger, D. (1990). *Individual and organizational response to the 1985 earthquake in Mexico City, Mexico.* http://udspace.udel.edu/handle/19716/2259

2. Endfield, G. H. (2012). The resilience and adaptive capacity of social-environmental systems in colonial Mexico. *Proceedings of the National Academy of Sciences, 109*(10), 3676–3681. https://doi.org/10.1073/pnas.1114831109

3. Layton, M. D., & Moreno, A. (2014). Philanthropy and social capital in Mexico. *International Journal of Nonprofit and Voluntary Sector Marketing, 19*(3), 209–219. https://doi.org/10.1002/nvsm.1498

4. Putnam, R. D. (1995). Tuning in, tuning out: The strange disappearance of social capital in America. *PS: Political Science & Politics, 28*(4), 664–683. https://doi.org/10.2307/420517

5. Cruz-Torres, C. E., & Martín del Campo-Ríos, J. (2022). Social capital in Mexico moderates the relationship of uncertainty and cooperation during the SARS-COV-2 pandemic. *Journal of Community Psychology, 50*(2), 1048–1059. https://doi.org/10.1002/jcop.22699

6. Elorriaga-Verplancken, F. R., Sandoval-Sierra, J., Paniagua-Mendoza, A., & Robles-Hernández, R. (2018). Seasonality and potential foraging grounds of migratory California sea lions from La Paz Bay, southern Gulf of California, Mexico. *Aquatic Mammals, 44*(1), 56–61. https://doi.org/10.1578/AM.44.1.2018.56

7. Costa, D., Weise, M., & Arnould, J. (2006). Potential influences of whaling on the status and trends of pinniped populations. In *Whales, whaling and ocean ecosystems* (pp. 344–359). University of California Press. http://hdl.handle.net/10536/DRO/DU:30008326

8. Salazar Pico, S. K. (2005). *Variación* temporal y espacial del espectro trófico del lobo marino de Galápagos (Zalophus wollebaeki). Maestría en Manejo de Recursos Marinos Thesis. https://www.repositoriodigital.ipn.mx/jspui/handle/123456789/14456

9. Páez-Rosas, D., & Aurioles-Gamboa, D. (2010). Alimentary niche partitioning in the Galapagos sea lion, Zalophus wollebaeki. *Marine Biology, 157*(12), 2769–2781. https://doi.org/10.1007/s00227-010-1535-0

10. *What are El Ni*ño and La Niña? (2023, August 24). National Oceanic and Atmospheric Administration. https://oceanservice.noaa.gov/facts/ninonina.html

11. Trenberth, K. E., & Hoar, T. J. (1996). The 1990-1995 El Niño-Southern Oscillation event: Longest on record. *Geophysical Research Letters, 23*(1), 57–60. https://doi.org/10.1029/95GL03602

12. Idyll, C. P. (1973). The anchovy crisis. *Scientific American, 228*(6), 22–29. http://www.jstor.org/stable/24923069

13. Páez-Rosas, D., Vaca, L., Pepolas, R., Wollocombe, R., de Roy, T., & Rivas-Torres, G. (2020). Hunting and cooperative foraging behavior of Galapagos sea lion: An attack to large pelagics. *Marine Mammal Science, 36*(1), 386–391. https://doi.org/10.1111/mms.12646

14. Poirier, F. E., Bellisari, A., & Haines, L. (1978). Functions of primate play behavior. In *Social play in primates* (pp. 143–168). Elsevier. https://doi.org/10.1016/B978-0-12-652750-6.50011-0

15. Gentry, R. L. (1974). The development of social behavior through play in the Steller sea lion. *American Zoologist, 14*(1), 391–403. https://doi.org/10.1093/icb/14.1.391

16. Poirier, F. E., & Smith, E. O. (1974). Socializing functions of primate play. *American Zoologist, 14*(1), 275–287. https://doi.org/10.1093/icb/14.1.275

17. deMenocal, P. B. (1995). Plio-Pleistocene African climate. *Science, 270*(5233), 53–59. https://doi.org/10.1126/science.270.5233.53

18. Buis, A. (2020, February 27). *Milankovitch (orbital) cycles and their role in Earth's climate.* NASA Global Climate Change. https://climate.nasa.gov/news/2948/milankovitch-orbital-cycles-and-their-role-in-earths-climate/

19. Lee, J., Shen, A., Fox-Kemper, B., & Ming, Y. (2017). Hemispheric sea ice distribution sets the glacial tempo. *Geophysical Research Letters, 44*(2), 1008–1014. https://doi.org/10.1002/2016GL071307

20. deMenocal, P. B. (2004). African climate change and faunal evolution during the Pliocene–Pleistocene. *Earth and Planetary Science Letters, 220*(1–2), 3–24. https://doi.org/10.1016/S0012-821X(04)00003-2

21. Bobe, R., Behrensmeyer, A. K., & Chapman, R. E. (2002). Faunal change, environmental variability and late Pliocene hominin evolution. *Journal of Human Evolution, 42*(4), 475–497. https://doi.org/10.1006/jhev.2001.0535

22. Dorey, F. (2020, October 2). *Homo habilis.* Australian Museum. https://australian.museum/learn/science/human-evolution/homo-habilis/

23. Leakey, L. S. B., Tobias, P. v., & Napier, J. R. (1964). A new species of the genus *Homo* from Olduvai Gorge. *Nature, 202*(4927), 7–9. https://doi.org/10.1038/202007a0

24. *Homo habilis.* (2022). Smithsonian Institution. https://humanorigins.si.edu/evidence/human-fossils/species/homo-habilis

25. Braun, D. R., Aldeias, V., Archer, W., Arrowsmith, J. R., Baraki, N., Campisano, C. J., Deino, A. L., DiMaggio, E. N., Dupont-Nivet, G., Engda, B., Feary, D. A., Garello, D. I., Kerfelew, Z., McPherron, S. P., Patterson, D. B., Reeves, J. S., Thompson, J. C., & Reed, K. E. (2019). Earliest known Oldowan artifacts at >2.58 Ma from Ledi-Geraru, Ethiopia, highlight early technological diversity. *Proceedings of the National Academy of Sciences, 116*(24), 11712–11717. https://doi.org/10.1073/pnas.1820177116

26. Smithsonian Institution. (2022). *Oldowan tools from Lokalalei, Kenya.* https://humanorigins.si.edu/evidence/behavior/stone-tools/early-stone-age-tools/oldowan-tools-lokalalei-kenya

27. Tobias, P. v. (1965). *Australopithecus, Homo habilis,* tool-using and tool-making. *The South African Archaeological Bulletin, 20*(80), 167. https://doi.org/10.2307/3887824

28. Plummer, T. (2004). Flaked stones and old bones: Biological and cultural evolution at the dawn of technology. *American Journal of Physical Anthropology, 125*(S39), 118–164. https://doi.org/10.1002/ajpa.20157

29. Wood, B. A., & Patterson, D. B. (2020). *Paranthropus* through the looking glass. *Proceedings of the National Academy of Sciences, 117*(38), 23202–23204. https://doi.org/10.1073/pnas.2016445117

30. Ungar, P. S., Grine, F. E., & Teaford, M. F. (2006). Diet in early *Homo*: A review of the evidence and a new model of adaptive versatility. *Annual Review of Anthropology, 35*(1), 209–228. https://doi.org/10.1146/annurev.anthro.35.081705.123153

31. Ungar, P. S. (2012). Dental evidence for the reconstruction of diet in African early *Homo. Current Anthropology, 53*(S6), S318–S329. https://doi.org/10.1086/666700

32. Potts, R., & Shipman, P. (1981). Cutmarks made by stone tools on bones from Olduvai Gorge, Tanzania. *Nature, 291*(5816), 577–580. https://doi.org/10.1038/291577a0

33. Bunn, H. T., & Ezzo, J. A. (1993). Hunting and scavenging by Plio-Pleistocene hominids: Nutritional constraints, archaeological patterns, and behavioural implications. *Journal of Archaeological Science, 20*(4), 365–398. https://doi.org/10.1006/jasc.1993.1023

34. Sebanz, N., Bekkering, H., & Knoblich, G. (2006). Joint action: Bodies and minds moving together. *Trends in Cognitive Sciences, 10*(2), 70–76. https://doi.org/10.1016/j.tics.2005.12.009

35. Dunbar, R. I. M. (1991). Functional significance of social grooming in primates. *Folia Primatologica, 57*(3), 121–131. https://doi.org/10.1159/000156574

36. Goosen, C. (1981). On the function of allogrooming in Old-World monkeys. In *Primate behavior and sociobiology: Selected papers (Part B) of the VIIIth Congress of the International Primatological Society, Florence, 7–12 July, 1980* (pp. 110–120). https://doi.org/10.1007/978-3-642-68254-4_16

37. Dunbar, R. I. M. (2022). Laughter and its role in the evolution of human social bonding. *Philosophical Transactions of the Royal Society B: Biological Sciences, 377*(1863). https://doi.org/10.1098/rstb.2021.0176

38. Lehmann, J., Korstjens, A. H., & Dunbar, R. I. M. (2007). Group size, grooming and social cohesion in primates. *Animal Behaviour, 74*(6), 1617–1629. https://doi.org/10.1016/j.anbehav.2006.10.025

CHAPTER 6

Through the Grapevine: Information

I.

The dew on the cockpit cushions was dry. Surrounding the boat, Butterfly Bay was glass.

"Good morning," my husband said, as he surfaced from the companionway, carrying his snorkel, mask, and fins.

"Morning," I said. Within 30 seconds, he splashed into the water.

I unrolled the chart of the Whitsundays on the cockpit table. As I did, a horseshoe-shaped beach near the bottom caught my eye. The name, Whitehaven Beach, sounded familiar to me. After our first night under attack by the bullets of wind, I liked the sound of any place called "haven." The anchorage closest to it was Chalkie's Beach. It looked like the fastest route there was around the northeastern tip of Hook Island and down the eastern, ocean-facing sides of Hook Island and Whitsunday Island. Or, you could go inside the shelter of the islands the whole way, but I figured that would be several hours longer.

My husband was still in the water when the 8:00 a.m. roll call came around. I went downstairs and switched on the radio. The voice on channel 82 was still on boats beginning with "A," so I tuned out while the others reported their plans for the day. Then, I heard the name of our boat. I quickly grabbed the microphone and repeated it.

"We're currently anchored in Butterfly Bay," I said, playing with the coiled microphone cable. "We plan to sail down the eastern side of Hook Island and Whitsunday Island and anchor tonight at Chalkie's Beach."

There was a moment of silence on the other end.

"Copy that," the man said. He paused again. "It can get choppy around Pinnacle Point when you have wind against tide, so you might have to take it slow." He went on, "It's meant to pick up to 25 knots later this morning. If it gets too rough on the eastern side of the islands, you should come through Hook Passage and down the western side."

"Copy that," I said, trying to sound as maritime as he did. I slid the microphone back into its slot. My cheeks felt hot, and I began to open the hatches. I got the sense that I was in school again. *Isn't this our charter?* I thought. *He's probably used to customers who don't want to spill their champagne. That's not us—we love adventure.* Then again, 25 knots of wind did sound like a lot.

Minutes later, my husband emerged from the water to tell me all about the funny-looking fish he had seen. He must have sensed in my eyes that my morning zen had blown away.

"What's wrong?" he asked. I told him what the man from the charter company said about going the other way due to the wind against tide at Pinnacle Point. I watched his face turn to stone and his eyes glaze over. I had seen this calcification happen many times before—he hated being told what to do a thousand times more than I did.

"Nah, I reckon we give it a go," he said. With that, he descended into the cabin in loud, resolute footsteps.

Within the hour, we had lifted anchor and were on our way. I took the helm first while my husband showered. We motored towards the infamous Pinnacle Point.

Approaching Pinnacle Point from the west, I could see the swell before I felt it. Visible from afar were the waves crashing against the rocks. They were a frothy white, like the thrashing suds inside a washing machine. I stepped forward to look up the mast at the wind vane. The meter had accelerated from the lazy rotation I saw earlier that morning to a briskly rushing wheel.

"Twenty-five knots," I repeated to myself.

The next thing I knew, the waves were there to greet me. They started hitting us side on. The boat began to rock from side to side. As we rocked, the mast swayed from left to right, left to right, like a mechanical metronome. The standing rigging and mast started to squeal under the tension. I dug in my heels and tightened my grip on the wheel.

I didn't know what to do, but I sensed there was way too much pressure on the rigging, and it might snap. I had to ease it somehow. I thought if I could take the waves at an angle rather than having them hit us side on, we might rock a little less. I tried adjusting our heading, pointing us due east and out to sea. Not knowing much about boat construction, I wondered how much stress the vessel could handle before the mast would break. I hadn't the first clue what to do if we lost the mast.

Changing our heading eased the rocking, but then we encountered a different problem. With the waves coming head on, the fiberglass bow of the boat began to slam the water's surface between each wave. BAM ... BAM ... BAM. The sound of the slamming was deafening. I was stuck: If

I turned back, the mast would resume its side-to-side rocking, but if we kept slamming against the water like we were, I could just picture the boat cracking in half, Titanic style.

My husband came up the companionway, fresh from his shower. From his expression, he seemed energized, rather than alarmed, by the commotion.

"Do you want to have a go?" I shouted, hoping for a reprieve.

"You seem to be doing alright," he said.

"Not really," I responded. "It's pretty rough."

He quickly took over the helm. Having surfed in much bigger waves as a teenager, he was rarely intimidated by swell.

"We're meant to be headed southeast," I told him, "but the waves keep hitting us side on, so I've been steering a bit more to the east." I pointed my arm in the direction of our bearing. "We can get back on course after we get through this rough patch."

Ten minutes later, and we were still not through the rough patch. I was down below, grateful to no longer be at the helm. But I could still hear the boat's brutal screeching. It sounded even louder because I was sitting just below the mast. I had popped a TravaCalm to help with my sea sickness, but my mind was racing. Here we were, headed farther and farther out to sea, where the waves would just get bigger and bigger.

I couldn't sit still any longer. I climbed up the companionway to see how my husband was doing. The calm confidence on his face was gone. I realized he didn't have a good plan for these waves either.

Scrambling my brain for what to do, I suddenly recalled what the man from the charter company had told me during the roll call. He said we should go slow and, if things got rough, duck in through Hook

Passage. The western side of the islands would be sheltered from the wind and waves.

"I have an idea," I said to my husband. His eyes clung to mine. I said, "The guy from the charter company said if it's too rough, we should come through Hook Passage."

I saw him cogitating.

"But that's going to take forever," he said. "This isn't that bad. We can ride it out."

I knew he would never choose to back away from a challenge, but I was insistent.

"The wind is just going to pick up," I said. "And we're headed out to sea. We can't stay on this bearing."

I didn't know much about conditions at sea around the Whitsundays, but the guy from the charter company obviously did. We didn't need to stick it out any longer to find out just how rough it could get.

Reluctantly, my husband conceded to the plan. He turned down the revs on the engine and pointed towards Hook Passage. Our progress slowed considerably, but at least the bow stopped slamming on the water.

We approached Hook Passage, swishing and sliding in the swell. As soon as we rounded the corner, the wind died off. The water returned to a glassy turquoise. Once again we were in the safe, tropical paradise of the Whitsunday Islands.

I caught my breath. My mind moved to what we would have done if the man from the charter company hadn't told us what to do. Not knowing, we might have stuck out the rough seas longer, which could have ended in a real disaster. Although I was loath to admit defeat, this time it definitely wasn't our own smarts that saved us. It was only because we heeded the advice of a more experienced sailor that we made it to safety.

II.

Sometimes we get through the diciest situations in life only because a tidbit of knowledge saves us. This is especially true at sea. Being on the ocean is one of the most dangerous things you can do with your life. Even with all the assistance of modern navigation and weather forecasting technologies, the mortality rate of commercial fishers is still over 29 times higher than the average among U.S. workers.[1] Nearly half of these deaths occur because of a vessel disaster, and another 30% occur when fishers fall overboard.[1] Put bluntly, there's nothing safe about the ocean.

There is a string of little port towns along the North Sea coast of the United Kingdom between Edinburgh and Newcastle upon Tyne, in a region called Northumberland. There the fishers use baited pots to catch European lobster (*Homarus gammarus*). Lobster is a high-value species, commercially speaking,[2] commanding a price of £18/kg in the United Kingdom (2021 prices).[3] But lobsters are difficult to find, hiding in rocky recesses in the coastal seafloor. It takes a lot of experience to know where to look.[2] This kind of experience is often lacking in the larger, more urban communities towards the south, where seasonal and part-time newcomers appear during lobster season from July to October, often with little or no fishing background. On the other hand, in the smaller, more rural ports in the north, fishing is undertaken by generations of the same family. The fishing families form part of close-knit communities,[2] which provide the perfect conditions for sharing experience.

Because of the ocean's unpredictability, expert knowledge about seafaring is extremely valuable. The salty, wise fisherman is not just a stereotype. Multiple studies have found that commercial fishers rely on their experiential knowledge of the ocean (i.e., their wisdom) and the knowledge of other fishers to make decisions on where to fish for their target species.[4] This knowledge is what helps them to mitigate some of the uncertainty in their line of work.[4, 5, 6, 7]

Many people consider knowledge to be an individual asset—it's something you possess and is unique to you. However, Rachel Turner at the University of the West Indies and her colleagues at Newcastle University were not convinced. They had seen growing evidence that fishers share information with each other through their social relationships. *If we were to reframe knowledge as a social asset rather than an individual asset*, they wondered, *how would knowledge relate to the success or failure of entire communities?*

In 2013, Turner and her colleagues journeyed to the North Sea to investigate. Their mission was to explore the relationship between information-sharing networks and fishing success in the Northumberland lobster fishing industry.[2]

Four port towns along the Northumberland coast were chosen as study sites. The southernmost city of Blyth had a population of around 37,000 (2001 census). The next town towards the north was Amble, with a population of around 6,000. Even further north was the village of Seahouses, with a population of about 1,800. And at the northernmost tip of the Northumberland coast was the community of Holy Island, with a mere 162 residents. The team took to asking questions of the fishers in each port. There was some territoriality among groups of fishers in

Northumberland, so for the purposes of their study, they considered each port to be a separate group or social network.[2]

Turner and her colleagues interviewed 90% of the fishers in the four towns. They reported using baited pots to capture lobster along the River Tyne and within 6 nautical miles of the shore. During peak season, a single fishing operation could set in excess of 2,000 pots. They targeted skippers of active potting vessels, because from their preliminary interviews, they learned that it was the skippers who were making the decisions about where to fish each day. They asked the skippers questions like, "With which other fishing vessels or skippers do you share information about fishing location or catch?" Any time another fisher or another vessel was mentioned, the researchers recorded this as a "social tie." Eventually, over the course of 41 interviews, they were able to add more and more ties, creating a web of ties for each port. The four webs represented the networks across which the fishers were sharing information.[2, 8]

Of the four ports, the northernmost community of Holy Island was the most cohesive network of fishers. Holy Island was also the smallest of the four ports and had the greatest number of social ties per population. Unlike in other ports, all fishers in Holy Island had more-or-less equal access to information; one person wasn't controlling the information.[2] Anyone who has ever lived in a small town can relate to the situation at Holy Island. When there is a small population, you tend to know everyone, and secrets travel quickly. The larger the population, as in the southern ports of Blyth and Amble, the less dense the social network tends to be. In these ports, it's impossible to maintain close relationships with everyone.

What is more interesting than the small-town cohesion, though, is how information sharing impacted the fishers' success in catching lobsters. Turner and her colleagues discovered that, across the four ports, the fishers who had more incoming information-sharing ties were likely to land more lobsters.[2] In other words, their social networks were contributing directly to their success in terms of cold, hard lobster. While this may not seem unusual (most of us already had a hunch that better connected individuals tend to be more successful), for our purpose of understanding the evolution of human cooperation, it pays to look at why this is so.

Learning by trial and error is the most basic form of animal learning. It works well most of the time for most animals. But trial and error is a costly form of learning. Sometimes we don't have the luxury of trialing multiple, different options and waiting around to see how they turn out. The lobster fishers in Northumberland couldn't waste time testing every fishing spot within 6 nautical miles of the coast to figure out where there were lobster. They needed a shortcut to that knowledge, or they wouldn't have been able to feed their families. So they asked a more experienced friend, "Hey, where have you found lobster lately?" and this friend sent them off in the right direction. In other words, they shortcut the lengthy process of trial and error by sharing information.

Social capital theory suggests that our social connections contribute to our success in life. But why? Some sociologists have theorized that people who have more social connections perform better because they have access to a broader knowledge base than people with fewer connections.[9, 10] If you add up all of the trial and error-gained experiences of each of your social connections, assuming they're willing to share them, then your knowledge is actually the sum of all of your friends' knowledge. For surviving harsh and dynamic environments, whether the ever-changing waters of the ocean or the changing African landscape, knowledge is often the only thing that can save us. It really helps to know people who already have that knowledge.

If you never leave the place where you were born, you can gain all the important knowledge you need about your environment through trial and error. But when it comes to venturing out into a new, uncertain, or unpredictable environment, like the ocean, trial and error is a costly and potentially dangerous strategy for acquiring knowledge. Communication with other members of our social group shortcuts the costly process of trial and error by pooling the knowledge of multiple individuals.

Communication is, by definition, the transfer of information between a sender and a receiver.[11] Humans do a lot of communicating with those in our social network, but so do other animals.[11] This indicates that there must have been a pretty strong evolutionary pressure for communication to evolve in social animals. Communication must have signified survival to us, as it does to the lobster fishers in Northumberland. For communication to have evolved in humans, it must have been an essential asset in the environment of our ancestors, allowing us to venture farther than we ever could on our own.

For the survival of humanity, trial and error alone didn't cut it. Our ancestors ventured around Africa, out of Africa, and across the globe. We needed a way of pooling our knowledge. We needed to share information.

III.

In the highly unpredictable environment of the ocean, fishers aren't the only ones who rely on sharing information to survive.

Ocean scientists Maurice Ewing and J. Worzel were aboard Research Vessel *Saluda* in the spring of 1944. They were in the deep ocean, off the shore of Massachusetts. From their vessel, they lowered a hydrophone—an underwater device that records sounds from all directions—to a depth of about 1,219 meters. At a designated time, another ship dropped a 2 kilogram explosive to detonate at the same depth. The ship that dropped the explosive was 1,448 kilometers away, a quarter of the way across the Atlantic. The pair of scientists, not knowing whether any sound of the explosion would reach them, listened attentively through the hydrophone.[12]

What Ewing and Worzel heard through the hydrophone was unmistakable. There was a pulse, and then another, and then another, with each pulse louder than the one before. The pulses climaxed in a sound that was "so sharp that it was impossible for the most unskilled observer to miss it."[13] What Ewing and Worzel had stumbled upon was the horizontal layer of the ocean at which low-frequency sound travels farthest. At depths of around 1,000 meters, sound waves can travel for thousands of kilometers, across entire ocean basins, without dissipating.[14] This special sound tunnel lying deep in the ocean later became known as the SOund Fixing And Ranging channel, or SOFAR channel.

The U.S. Navy was quick to tap into the potential of the SOFAR channel. They began using it during WWII to listen for downed aircraft and sinking ships. At the height of the Cold War, the U.S. Navy placed entire webs of hydrophones on the ocean floor. The hydrophone arrays were connected by underwater cables to listening stations. Under the project name Jezebel, the U.S. Navy began using this system to detect and plot the position of Soviet submarines.[13] It was only then that they realized they were not the only ones tapping into the SOFAR channel.

The Navy sonar technicians began to hear strange clicking noises. They were coming through the hydrophones at the very low frequency of 20 hertz. Perhaps there was something wrong with the amplifiers on the ship? No, the amplifiers were fine. They spoke with scientists, who suggested that perhaps the sounds were from seismic activity deep within the Earth or breaking surf on continental shorelines. But when these sources were investigated, they didn't explain the loud 20 hertz signals either. Some of the more superstitious Navy sailors rumored that the sounds were coming from a creature deep within the ocean itself, whom they called the "Jezebel Monster."[13]

The mysterious clicks coming from the ocean were clearly not a straightforward research problem. For this reason, it took a pair of young scientists to fall in love and go fishing for answers before the Jezebel Monster case could be resolved.

William Schevill was a young man with an unquenchable thirst for paleontology. After graduating from Harvard in the early 1930s, he began work at Harvard's Museum of Comparative Zoology. His research took him all over the world in search of fossils. Schevill's responsibilities at Harvard grew with time: He became the Associate Curator of Invertebrate Paleontology and later the Museum's Librarian. It was through his duties at the Museum Library that he came to meet another budding scientist, Barbara Lawrence. Lawrence was studying under the Museum's Director and the Curator of Mammals, and like Schevill, she was undertaking her own research projects and mounting her own expeditions. Between her trips to the Philippines and Sumatra, she spent long hours back at Harvard poring over zoological texts in the Museum Library. It was there that she first set eyes on Schevill.[15]

Schevill and Lawrence were wed in 1938, but with the start of WWII,

Schevill was abruptly pulled into matters of national defense. Given his scientific background, he was placed with a small group of Allied submariners who were attempting to locate German U-boats by emitting long-distance sounds through the recently discovered SOFAR channel.[15] It was during his time on the Allied submarines that Schevill was first introduced to the Jezebel Monster. The Navy sonar technicians pointed out the many strange sounds coming through their hydrophones, which seemed to be emanating from the deep. Schevill didn't know it at the time, but he would dedicate the rest of his career to investigating this so-called monster.[16]

Schevill's dinnertime stories of the strange sounds coming through the hydrophones must have captivated his wife, because after the war Lawrence joined her husband in investigating the Jezebel Monster. For centuries, sailors had been recounting tales of strange and haunting sounds coming from the sea through the hulls of their boats. So Schevill and Lawrence sought information from as many sailors as they could find, believing that their collective years of experience at sea might shed some light on the situation. Sailors reported to Schevill and Lawrence that they heard the sounds only on rare occasions, when whales had been spotted in the vicinity of their boats. The entire hull of the boat would begin to vibrate, acting as an amplifier of the sound waves from the water.[17] Schevill and Lawrence were intrigued by the sailors' theory that the underwater sounds could be coming from whales. Being good scientists, though, they needed to observe it for themselves.

Observing whales in the wild in the act of producing sound is fraught with difficulties. Human ears can distinguish direction in air, but not in water, so if you're under water and you hear a sound, it's impossible to know where to look to find the source of the sound. What's more, whales don't open their mouths to make sound, so to an observer, they appear no different if they are vocalizing or are silent.[17] Therefore, the only way that Schevill and Lawrence could test whether whales were responsible for the sounds reported by the sailors and Navy technicians was to find a place where they could clearly catch a whale in the act of making sound.

Shortly after the war, Schevill and Lawrence traveled to Florida. Marine Studios (now Marineland) was located just south of St. Augustine and was one of Florida's main tourist attractions at the time. It had been built in 1938 and was a popular place among Lost Generation writers, including Ernest Hemingway. Nevertheless, Schevill and Lawrence were not at Marine Studios for the tourist experience. They came for one thing and one thing only: to hear a whale sing.[15]

Schevill and Lawrence chose Marine Studios because they had come across an article written by Arthur McBride, Marine Studios' curator, in which he reported hearing various whistling sounds coming from the bottlenose dolphins (*Tursiops truncatus*) in the park.[18] McBride reported that when the dolphins were first introduced to their tank at Marine Studios, he noticed sharp whistling sounds echoing throughout the corridors surrounding the tank. Once the dolphins had acclimated to their environment, however, the whistling died off somewhat and seemed to be used by the dolphins only for communication with each other. He reported that the dolphins also whistled excitedly when they were chasing live fish. In his article, McBride surmised, "Apparently the animals used variations in this whistling sound to indicate different emotional states."[18]

Most people don't think of them as whales, but dolphins are actually a kind of toothed whale. Reading McBride's account of dolphin vocalizations, Schevill and Lawrence became eager to observe the phenomenon for themselves.

Upon their arrival at Marine Studios, Schevill and Lawrence were guided to the circular dolphin tank. As they approached, they could hear whistles and clicks reverberating throughout the corridors. The dolphin tank was lined with glass portholes,[18] providing visitors with "a window in the sea."[19] Schevill and Lawrence peered through the portholes into the iridescent waters that had been pumped in from the ocean. They laid eyes on the gracefully moving pearl-gray bodies of the bottlenose dolphins. What they witnessed was astonishing—it was just as McBride had reported. Although they possessed no vocal cords, the dolphins were somehow capable of producing an array of sounds underwater without letting any air escape.[15]

After their tantalizing glimpse at toothed whale vocalizations at Marine Studios, Schevill and Lawrence journeyed into the wild to see whether they could observe the same phenomenon. They plopped their hydrophone into the frigid waters of Quebec's Saguenay River, where beluga whales (*Delphinapterus leucas*) were frequently spotted. Not long after, they began to hear the following:

> *High-pitched resonant whistles and squeals, varied with ticking and clucking sounds slightly reminiscent of a string orchestra tuning up, as well as mewing and occasional chirps. Some of the sounds were bell-like, and a few rather resembled an echo sounder. Occasionally the calls would suggest a crowd of children shouting in the distance. At times there were sharp reports, somewhat like a blow with a split bat or a slap on the water, although nothing could be seen to be striking the water in any way.*[20]

Schevill and Lawrence had discovered an entire symphony underwater. Throughout their time on the Saguenay River, they were able to establish a correlation between the sounds they heard and the abundance of beluga whales passing through the river at the time, producing the first scientific finding of vocalization by wild whales. Beluga whales, the chatter birds they turned out to be, became known as "sea canaries."[20]

Throughout their lifetimes, Schevill and Lawrence went on to record the underwater vocalizations of more than 60 species of marine mammals.[20] New models of hydrophone needed to be invented to receive the very low-frequency sounds the animals produced. It was during their recordings that Schevill finally heard the same loud, 20 hertz pulses that he had heard in the SOFAR channel while working in the Navy. Some of the pulses were over 80 decibels, which is the same volume as a freight train passing right in front of you.[21] According to Schevill's colleague Bill Watkins, "The relative precision of pulse repetition, the sinewave character of the sounds and the long repetitive patterns, all seemed to indicate man-made signals."[21] Such pulses certainly sounded like they could have been the Soviets, but being surrounded by fin whales (*Balaenoptera physalus*) in Cape Cod, Massachusetts, Schevill and Watkins finally concluded that these 20 hertz pulses were, in fact, coming from fin whales.[21]

The Jezebel Monster mystery was solved at last. And it was a monster of sorts—the fin whale is one of the largest whale species on the planet, second only to the blue whale.[22] It appears that the Americans weren't the only ones taking advantage of the sound travel properties of the deep ocean to communicate. Fin whales were too.

Oceanographer Douglas Webb and whale biologist Roger Payne were the first to theorize that fin whales use the SOFAR channel to communicate with each other across thousands of miles in the deep waters of the ocean.[23] They reasoned that the SOFAR channel would help to propagate the low-frequency sounds produced by fin whales.[17] In the 1970s and '80s, many marine biologists were skeptical of Webb and Payne's theory of long-range communication by whales. But in the early '90s, Payne's former student Christopher Clark produced experimental data substantiating the theory. Clark reported that while he was listening through a hydrophone placed just off Bermuda, he could hear a whale singing all the way in Ireland.[17] When he went on to create a map of how far the whale sounds were propagating, it illuminated the entire ocean.[24]

There could be many reasons that fin whales choose to stay in touch across the ocean. Many social animals use vocalizations to communicate with each other so that they can select a mate, avoid predators, parent offspring, find food, or coordinate group activities.[11] Fin whales are big animals, and throughout their long lifespan, they require a lot of resources to survive and reproduce. The main food source of fin whales is krill, a small crustacean similar to shrimp.[22] Krill feed on phytoplankton, which bloom each winter in the nutrient-rich waters around the Arctic and Antarctica. Each year, fin whales leave their tropical calving grounds and journey to the poles so that they can get their fill of the phytoplankton-fed krill. Phytoplankton blooms do not occur in the same spot each year, however. Blooms are heavily dependent on weather anomalies. Therefore, it would be rare for a fin whale to find krill in the same spot 2 years in a row.

Given the variability in where krill hang around, it may benefit whales to spread out and communicate with each other once they've found food. Kind of like, *You go this way, I'll go that way—let me know if you find anything.* This strategy of sharing information about good feeding opportunities may increase the whole herd's chances of finding food each year.[17, 25] The advantages of looking for food as a group have been shown in many animals, with larger group sizes finding food more quickly.[26] In contrast, searching for food on one's own uses a lot of time and energy, and it puts individuals at risk by exposing them to predators. By cooperating to find food sources and sharing information about them, individuals can more readily ascertain the location of food, the amount of food available, the suitability of the location, or whether there are any predators nearby.[26]

Sharing information about feeding sites is particularly useful if resources are sparse.[27] This was certainly the case with the lobster in Northumberland. It may also be the case with the krill hunted by fin whales. In search of valuable resources, both fishers and whales have developed ways of sharing information with each other to increase their likelihood of a catch. This strategy of sharing information could have also helped our ancestors to increase their "catch," especially as they ventured farther.

IV.

For the majority of hominin evolution, our ancestors had little need for long-distance travel. Their fossilized remains and tools suggest that prior to 1.77 mya, they generally stayed close to their home bases. They lived in the margins of lakes, often traveling no more than 12 kilometers in search of stone to make tools.[28] But for some reason, around 1.77 mya, they began to venture.

The first species of hominin who is thought to have ventured far and wide is *Homo erectus* (meaning "upright man" in Latin). The travels of this upright-walking species were uncovered in 1891 when a surgeon in the Dutch army stationed on the island of Java came across a well-preserved skullcap of an individual who looked to be an intermediate form between humans and apes. The surgeon, Eugène Dubois, named the species *Pithecanthropus erectus*, or "erect ape-man." Anthropologists later recognized that the fossils belonged to the same genus as humans, so the individual was assigned to the species *Homo erectus* and was given the nickname "Java Man."[29]

The discovery of Java Man opened a whole world of questions about human prehistory, the most obvious ones being, *How and why did* Homo erectus *leave Africa?*

The *how* question is the more straightforward one to answer. We can see from the anatomy of *Homo erectus* that this species was much better adapted to walking long distances than any hominin who had come before. Matching proximal and distal lower limb bones found at Dmanisi, Georgia, indicate that *Homo erectus* stood around 145–166 cm tall and weighed 40–50 kg. Limb proportions of *Homo erectus* were similar to those of modern humans, with much longer legs than earlier hominins had. Given that for land-based animals, the cost of transport decreases as the back legs get longer, *Homo erectus* likely had better locomotor energy efficiency than other hominin species who had come before. The large vertebrae of *Homo erectus*, coupled with a big toe that stood separate from the other toes and a high arch of the foot, allowed *Homo erectus* to walk and run as efficiently as modern humans.[30] All of these anatomical changes meant our ancestors could wander farther than ever before.

The question of *why* our ancestors left Africa is somewhat of a mystery. With *Homo erectus* remains scattered across Africa, Georgia,

Israel, China, and Indonesia, it's clear that this was a species capable of adapting to a variety of climates. But what could have possibly motivated our ancestors to travel so far?

Looking at the geological record, it appears that food sources for *Homo erectus* became increasingly sporadic in Africa around 1.9–1.77 mya. Sediments from the Turkana Basin in northern Kenya indicate higher-than-usual wet–dry variability during this period, with frequent fluctuations in the lake level and vegetation.[31] There was a major environmental shift at least every 37,000 years. What started as wet, marshy woodlands conditions gave way to arid, open plains, which was followed by a return to moist woodlands, before turning to very dry conditions with barely any vegetation.[32] Evidence from the bedrock indicates that the African continent was being broken up by tectonic movement, creating new rifts and mountain ranges.[33] It's likely that the permanent lakes frequented by our early ancestors began to break apart and disappear. Rivers took completely new paths. Populations of animals in East Africa were booming one minute, and the next minute, dwindling.[31]

Like fin whales in search of krill, our ancestors on the African continent would have been forced to navigate a mosaic of new habitats in search of food. It is possible that some populations of hominins overcame the unpredictability of food in their environment by expanding their geographic range. There is evidence that from around 1.77 mya, hominins began to occupy sites outside of lake margins for the first time, and they developed a wider-ranging, more variable lifestyle.[28] Some of our ancestors became so adept at occupying foreign environments that soon they were no longer confined to Africa.

Shortly after 2 mya, several species of large African mammals dispersed out of Africa and into the Middle East and western Asia.[28] It has been suggested that some groups of *Homo erectus* might have followed these mammals, as their primary source of food. Waves of hominin dispersals out of Africa are thought to have happened sporadically and in multiple directions. The oldest hominin remains outside of Africa are at Dmanisi, Georgia, from about 1.77 mya.[34] Older stone tools have been discovered in Jordan (2.5 mya) and China (2.1 mya), suggesting that hominins might have arrived in Asia even earlier.[35]

Fossils suggest that by around 1.66 mya, *Homo erectus* had dispersed throughout East Asia, occupying a latitudinal range from 7°S to 40°N.[28] There they inhabited temperate woodlands and grasslands as well as tropical and subtropical wooded landscapes. These new landscapes were under the influence of Asian monsoonal conditions and had plant and animal communities unlike anything *Homo erectus* had experienced in East Africa. Evidently, groups of *Homo erectus* had enough behavioral flexibility to adjust to this wide variety of habitats.[28]

One thing that might have helped *Homo erectus* was the size of their brains. In the 1 million years prior to the emergence of *Homo erectus*, hominin brains doubled in size.[36] Although we'll never know the exact causes of this doubling, it is likely that greater behavioral flexibility came along with larger brains. For example, larger brains might have made more advanced technological innovation possible. Archaeological findings suggest that *Homo erectus* were the first hominins to use Acheulean percussion flaking technology to make two-sided, pointed stone hand axes (around 1.7 mya).[37] A broader set of tools might have helped *Homo erectus* to survive in a broader range of climates. Their ability to innovate would have also aided their expansion into the colder northern latitudes of Eurasia.[29]

The enhanced behavioral flexibility of *Homo erectus* may explain why they were the longest-spanning hominin species to have ever lived. Unlike *Homo sapiens*, who have only been around for the last 300,000 years, *Homo erectus* first appeared in the fossil record around 2 mya in South Africa and was still around as recently as 108,000 years ago, having spread all over Asia and Africa.[38]

More cooperative forms of social interaction would have been critical to our ancestors' expansion out of Africa. To expand their range and find food in a variety of new habitats, it's possible that, like fin whales, *Homo erectus* needed to come up with a better system for acquiring information about their environment. Visual signs left by other animals in the environment would have constituted valuable information to our ancestors. Animal paw prints or hoof prints, disrupted ground foliage, broken sticks, and scratch marks are all subtle clues that someone else has been around. Perhaps *Homo erectus*, or even earlier hominins, learned through exploration and trial and error to associate certain signs with certain animals and circumstances. Historian Martin Stuart-Fox suggests that this could have been the origin of causal cognition in hominins.[39] In other words, as our ancestors wandered, they noticed visual signs, and they began to put two and two together.

In terms of a system for acquiring information, Northumberland fishers could testify that trial and error is a costly one. It would have taken ages for an individual to learn, by association alone, which paw print in the dirt or scratch mark on a tree was linked with which animal. What's more, *Homo erectus* populations were usually on the move. Like the whales in search of krill, their subpopulations would have traveled long distances in search of resources, moving across fractured zones of habitat. It would have taken an individual a lifetime just to understand the plethora of environmental signs around them, and by that time, much of their environment would have changed.

The best way for our ancestors to short-cut the expensive process of learning by trial and error would have been to share information with each other somehow. To achieve this level of sharing, they would have needed a common system of meaning. Some knowledge could have been acquired by imitation at first, but as evolutionary biologist Kevin Laland and colleagues have shown experimentally for tool manufacture, learning by imitation is nowhere near as effective as learning through verbally transmitted information.[40] Anthropologist Robert Boyd and colleagues have suggested that the complex methods of acquiring food used by members of our *Homo* genus—methods relying on a suite of specialized tools, detailed bodies of local knowledge, and complex cooperative systems—were only possible because we developed the ability to learn from others.[41] This claim is supported by studies of childhood learning in modern hunter-gatherer societies. Almost everything that children in these societies know about how to navigate, hunt animals, and find food is taught to them by their parents and elders over the course of their development.[42] This emphasizes that, unlike many animals, humans are not born knowing how to feed ourselves. We must learn how to survive from others.

Our hominin system of social learning likely began with very basic vocalizations. Like the sentinel meerkat on the mound, who has specific vocal calls for specific predators, our ancestors could have developed their own vocal system for referring to things in their environment. It could have started simply by pointing to an animal track and making a grunt,[43] and other tribe members coming to learn to associate that specific grunt with that specific animal. Eventually, through the repeated pairing of the two, when anyone in the group made that specific grunt, the image of that animal would have reliably appeared in everyone's mind. The main difference between the meerkats' system and our ancestors' system (as far as we know) is that the meerkat makes the call as she is looking at the predator, whereas our ancestors, who traveled and presumably tracked animals over long distances, would have needed language that could refer to animals and places that were familiar to them but not always in sight. In other words, we needed a language capable of transmitting mental representations.[44]

Rudimentary vocal calls associated with mental images would have been sufficient for most of the information that our ancestors needed. Going beyond these referential calls, though, our ancestors would have eventually needed specific sounds or words that would allow them to communicate cause and effect.[39] For example, "*A* eats us, but we eat *B*." Because they ventured so far and wide, they would have needed to explain spatial maps to each other and where certain things were located in relation to other things, which would have been critical for navigation. Grammar allows for action and relationships between concepts, which allows us to tell stories of cause and effect and give directions. Developing grammatical language might have meant that our ancestors could venture more widely and share their learnings with the group. It also meant that information could be reliably passed down from one generation to another.

It's possible that vocal communication was going on for hundreds of thousands of years before there was any fossil evidence of it, making it difficult to know when language actually began. It has been suggested that the anatomical potential to produce and perceive different vowel sounds emerged as long as 27 million years ago, at the time of our last common ancestor with Old World monkeys.[45] From there, hominin speech most likely developed over multiple, long stages.[45] The Acheulean tool-making technology used by *Homo erectus* to develop retouched hand-axes, cleavers, picks, and knives required much greater technical skill and planning compared with older forms of tool manufacture. This level of complexity, combined with fact that tools were manufactured in this same way for around 700,000 years with little variation, indicates that our *Homo erectus* ancestors were sharing some tradition of rules and techniques as long as 1.7 mya.[46] By 1 mya, our ancestors had ventured by boat to the island of Flores, Indonesia. It has been argued that by this stage, our ancestors required language both to build their boats and to plan their crossings of open bodies of water.[46] Hence, there is reason to believe (although, of course, not definitive proof) that language began with *Homo erectus*, if not earlier.

Charred bones and burnt dirt provide some of the first more-of-less definitive archeological evidence that our ancestors were sharing information across long distances. These and other signs of repeated fire use have been found at a few isolated sites from 800,000 years ago.[47] From around 400,000 years ago, the archaeological record is full of what seem to be hominin-made fires, appearing at sites across Europe, the Middle East, Africa, and Asia.[48] The finding that multiple hominin subpopulations scattered across cold and warm climates started using fire regularly around 400,000 years ago strongly suggests that the technology of controlling fire had a single point of origin.[48] In other words, rather than each group of hominins inventing fire technology independently, one group most likely discovered a method of starting and controlling fires, and they passed this information on to other groups with whom they came into contact.

As hominin groups traveled and interacted, the innovation of controlled fire spread throughout Africa and Eurasia.[48] There is evidence of gene flow among populations of ancient hominins,[49] so it is not unreasonable to assume that they learned from each other during their interactions, allowing a diffusion of innovations over space and time. This accelerated form of social learning, known as cultural transmission,[50] would have been aided greatly by early forms of language, especially if some individuals were versed in multiple languages and could act as knowledge brokers among the various groups. In reality, it's not that different to the cultural transmission of good ideas and knowledge that still happens among the human population today.

Faced with new environments, our ancestors would have relied heavily on the information shared through their cooperative networks to survive. Had their environment been constant and had they never ventured out of it, *Homo erectus* wouldn't have needed a more sophisticated system for sharing knowledge. But like the fin whales traveling across the ocean, our ancestors did venture. They ventured out of Africa and across Eurasia, taking their knowledge of the environment with them. And because our ancestors developed such an effective way of exchanging information, they eventually took their knowledge around the world.

CHAPTER 6 REFERENCES

Further Reading

Kenneally, C. (2018, September 1). The cultural origins of language. *Scientific American.* https://www.scientificamerican.com/article/the-cultural-origins-of-language/

References

1. National Institute for Occupational Safety and Health. (2023, July 13). *Regions: National overview*. Centers for Disease Control and Prevention. https://www.cdc.gov/niosh/topics/fishing/nationaloverview.html

2. Turner, R. A., Polunin, N. V. C., & Stead, S. M. (2014). Social networks and fishers' behavior: exploring the links between information flow and fishing success in the Northumberland lobster fishery. *Ecology and Society, 19*(2). http://www.jstor.org/stable/26269547

3. UK Government. (2022). *UK Sea Fisheries Statistics 2021.* https://www.gov.uk/government/statistics/uk-sea-fisheries-annual-statistics-report-2021

4. Johannes, R. E., Freeman, M. M. R., & Hamilton, R. J. (2008). Ignore fishers' knowledge and miss the boat. *Fish and Fisheries, 1*(3), 257–271. https://doi.org/10.1111/j.1467-2979.2000.00019.x

5. Salas, S., & Gaertner, D. (2004). The behavioural dynamics of fishers: Management implications. *Fish and Fisheries, 5*(2), 153–167. https://doi.org/10.1111/j.1467-2979.2004.00146.x

6. Grant, S., & Berkes, F. (2007). Fisher knowledge as expert system: A case from the longline fishery of Grenada, the Eastern Caribbean. *Fisheries Research, 84*(2), 162–170. https://doi.org/10.1016/j.fishres.2006.10.012

7. Marchal, P., Andersen, B., Bromley, D., Iriondo, A., Mahévas, S., Quirijns, F., Rackham, B., Santurtún, M., Tien, N., & Ulrich, C. (2006). Improving the definition of fishing effort for important European fleets by accounting for the skipper effect. *Canadian Journal of Fisheries and Aquatic Sciences, 63*(3), 510–533. https://doi.org/10.1139/f05-238

8. Scott, J. (2012). *What is social network analysis?* Bloomsbury Academic. https://library.oapen.org/bitstream/handle/20.500.12657/58730/1/9781849668200.pdf

9. Crona, B., & Bodin, Ö. (2006). What you know is who you know? Communication patterns among resource users as a prerequisite for co-management. *Ecology and Society, 11*(2), 7. http://www.ecologyandsociety.org/vol11/iss2/art7/

10. Carlsson, L., & Sandström, A. (2008). Network governance of the commons. *International Journal of the Commons, 2*(1), 33–54. https://www.jstor.org/stable/26522989

11. Simmons, A. M. (2003). Perspectives and progress in animal acoustic communication. In A.M. Simmons, R.R. Fay, & A.N. Popper (Eds.), *Acoustic communication. Springer handbook of auditory research* (Vol. 16, pp. 1–14). Springer. https://doi.org/10.1007/0-387-22762-8_1

12. Ewing, M., & Worzel, J. L. (1948). Long-range sound transmission. In J. L. Worzel, M. Ewing, & C. L. Pekeris (Eds.), *GSA memoirs: Propagation of sound in the ocean* (pp. 1–32). Geological Society of America. https://doi.org/10.1130/MEM27-3-p1

13. University of Rhode Island, & Inner Space Center. (2021). *History of the SOFAR channel.* Discovery of Sound in the Sea. https://dosits.org/science/movement/sofar-channel/history-of-the-sofar-channel/

14. NOAA Ocean Explorer. (2022, August 26). *SOFAR.* https://oceanexplorer.noaa.gov/explorations/sound01/background/acoustics/media/sofar.html

15. Backus, R. H., Bumpus, D., Lawrence, B., Norris, K. S., Ray, C. E., Ray, G. C., Twiss, J. R., & Watkins., W. A. (1995). William Edward Schevill 1906-1994. *Marine Mammal Science, 11*(3), 416–419. https://doi.org/10.1111/j.1748-7692.1995.tb00300.x

16. Rolfe, W. D. I. (2012). William Edward Schevill: Palaeontologist, librarian, cetacean biologist. *Archives of Natural History, 39*(1), 162–164. https://doi.org/10.3366/anh.2012.0069

17. Payne, R. (1995). *Among whales.* Scribner.

18. McBride, A. F. (1940). Meet Mister Porpoise. *Natural History, 1*, 16–29. https://www.yumpu.com/en/document/view/11468304/meet-mr-porpoise-ceta-base

19. Hill, R. N. (1956). *Window in the sea.* Rinehart. https://books.google.com/books/about/Window_in_the_Sea.html?id=SPseAAAAMAAJ

20. Schevill, W. E., & Lawrence, B. (1949). Underwater listening to the white porpoise (*Delphinapterus leucas*). *Science, 109*(2824), 143–144. https://doi.org/10.1126/science.109.2824.143

21. Watkins, W. A. (1981). Activities and underwater sounds of fin whales. *Sci. Rep. Whales Res. Inst., 33*, 83–117. https://www.icrwhale.org/pdf/SC03383-117.pdf

22. NOAA Fisheries. (n.d.). *Fin whale.* Retrieved November 17, 2023, from https://www.fisheries.noaa.gov/species/fin-whale

23. Payne, R., & Webb, D. (1971). Orientation by means of long range acoustic signaling in baleen whales. *Annals of the New York Academy of Sciences, 188*(1), 110–141. https://doi.org/10.1111/j.1749-6632.1971.tb13093.x

24. Bentley, M. (2005, February 28). Unweaving the song of whales. *BBC News.* http://news.bbc.co.uk/2/hi/science/nature/4297531.stm

25. Visser, F., Miller, P. J. O., Antunes, R. N., Oudejans, M. G., Mackenzie, M. L., Aoki, K., Lam, F.-P. A., Kvadsheim, P. H., Huisman, J., & Tyack, P. L. (2014). The social context of individual foraging behaviour in long-finned pilot whales (*Globicephala melas*). *Behaviour, 151*(10), 1453–1477. https://doi.org/10.1163/1568539X-00003195

26. Majolo, B., & Huang, P. (2022). Group living. In *Encyclopedia of animal cognition and behavior* (pp. 3021–3032). Springer International Publishing. https://link.springer.com/content/pdf/10.1007/978-3-319-55065-7_1865.pdf

27. Sueur, C., King, A. J., Conradt, L., Kerth, G., Lusseau, D., Mettke-Hofmann, C., Schaffner, C. M., Williams, L., Zinner, D., & Aureli, F. (2011). Collective decision-making and fission–fusion dynamics: A conceptual framework. *Oikos, 120*(11), 1608–1617. https://doi.org/10.1111/j.1600-0706.2011.19685.x

28. Potts, R., & Teague, R. (2010). Behavioral and environmental background to 'Out-of-Africa I' and the arrival of *Homo erectus* in East Asia. In J. G. Fleagle, J. J. Shea, F. E. Grine, A. L. Baden, & R. E. Leakey (Eds.), *Out of Africa I: The first hominin colonization of Eurasia* (pp. 67–85). Springer. https://doi.org/10.1007/978-90-481-9036-2_5

29. Smithsonian Institution. (2022). *Homo erectus.* https://humanorigins.si.edu/evidence/human-fossils/species/homo-erectus

30. Lordkipanidze, D., Jashashvili, T., Vekua, A., de León, M. S. P., Zollikofer, C. P. E., Rightmire, G. P., Pontzer, H., Ferring, R., Oms, O., Tappen, M., Bukhsianidze, M., Agusti, J., Kahlke, R., Kiladze, G., Martinez-Navarro, B., Mouskhelishvili, A., Nioradze, M., & Rook, L. (2007). Postcranial evidence from early *Homo* from Dmanisi, Georgia. *Nature, 449*(7160), 305–310. https://doi.org/10.1038/nature06134

31. Potts, R., & Faith, J. T. (2015). Alternating high and low climate variability: The context of natural selection and speciation in Plio-Pleistocene hominin evolution. *Journal of Human Evolution, 87,* 5–20. https://doi.org/10.1016/j.jhevol.2015.06.014

32. Hay, R. L. (1976). *Geology of the Olduvai Gorge.* University of California Press. https://doi.org/10.1525/9780520334229

33. Elton, S. (2008). The environmental context of human evolutionary history in Eurasia and Africa. *Journal of Anatomy, 212*(4), 377–393. https://doi.org/10.1111/j.1469-7580.2008.00872.x

34. Prat, S. (2018). First hominin settlements out of Africa. Tempo and dispersal mode: Review and perspectives. *Comptes Rendus Palevol, 17*(1–2), 6–16. https://doi.org/10.1016/j.crpv.2016.04.009

35. Scardia, G., Neves, W. A., Tattersall, I., & Blumrich, L. (2021). What kind of hominin first left Africa? *Evolutionary Anthropology: Issues, News, and Reviews, 30*(2), 122–127. https://doi.org/10.1002/evan.21863

36. Meyer, M. R. (2016). The spinal cord in hominin evolution. In *Encyclopedia of life sciences* (pp. 1–6). Wiley. https://doi.org/10.1002/9780470015902.a0027058

37. Diez-Martín, F., Sánchez Yustos, P., Uribelarrea, D., Baquedano, E., Mark, D. F., Mabulla, A., Fraile, C., Duque, J., Díaz, I., Pérez-González, A., Yravedra, J., Egeland, C. P., Organista, E., & Domínguez-Rodrigo, M. (2015). The origin of the Acheulean: The 1.7 million-year-old site of FLK West, Olduvai Gorge (Tanzania). *Scientific Reports, 5*(1), 17839. https://doi.org/10.1038/srep17839

38. Glausiusz, J. (2021, October 19). What drove *Homo erectus* out of Africa? *Smithsonian Magazine.* https://www.smithsonianmag.com/science-nature/what-drove-homo-erectus-out-of-africa-180978881/

39. Stuart-Fox, M. (2015). The origins of causal cognition in early hominins. *Biology & Philosophy, 30*(2), 247–266. https://doi.org/10.1007/s10539-014-9462-y

40. Morgan, T. J. H., Uomini, N. T., Rendell, L. E., Chouinard-Thuly, L., Street, S. E., Lewis, H. M., Cross, C. P., Evans, C., Kearney, R., de la Torre, I., Whiten, A., & Laland, K. N. (2015). Experimental evidence for the co-evolution of hominin tool-making teaching and language. *Nature Communications, 6*(1), 6029. https://doi.org/10.1038/ncomms7029

41. Boyd, R., Richerson, P. J., & Henrich, J. (2011). The cultural niche: Why social learning is essential for human adaptation. *Proceedings of the National Academy of Sciences, 108*(supplement_2), 10918–10925. https://doi.org/10.1073/pnas.1100290108

42. Scalise Sugiyama, M. (2011). The forager oral tradition and the evolution of prolonged juvenility. *Frontiers in Psychology, 2*. https://doi.org/10.3389/fpsyg.2011.00133

43. McCune, L. (2021). Laryngeal vocalizations: Grunts as a gateway to language? *Lingua, 264*, 103161. https://doi.org/10.1016/j.lingua.2021.103161

44. Scalise-Sugiyama, M. (2016). Gossip and grooming hypothesis. In *Encyclopedia of evolutionary psychological science* (pp. 1–8). Springer International Publishing. https://doi.org/10.1007/978-3-319-16999-6_3314-1

45. Boë, L.-J., Sawallis, T. R., Fagot, J., Badin, P., Barbier, G., Captier, G., Ménard, L., Heim, J.-L., & Schwartz, J.-L. (2019). Which way to the dawn of speech?: Reanalyzing half a century of debates and data in light of speech science. *Science Advances, 5*(12). https://doi.org/10.1126/sciadv.aaw3916

46. Barham, L., & Everett, D. (2021). Semiotics and the origin of language in the Lower Palaeolithic. *Journal of Archaeological Method and Theory, 28*(2), 535–579. https://doi.org/10.1007/s10816-020-09480-9

47. Sandgathe, D. M. (2017). Identifying and describing pattern and process in the evolution of hominin use of fire. *Current Anthropology, 58*(S16), S360–S370. https://doi.org/10.1086/691459

48. MacDonald, K., Scherjon, F., van Veen, E., Vaesen, K., & Roebroeks, W. (2021). Middle Pleistocene fire use: The first signal of widespread cultural diffusion in human evolution. *Proceedings of the National Academy of Sciences, 118*(31). https://doi.org/10.1073/pnas.2101108118

49. Hubisz, M. J., Williams, A. L., & Siepel, A. (2020). Mapping gene flow between ancient hominins through demography-aware inference of the ancestral recombination graph. *PLOS Genetics, 16*(8), e1008895. https://doi.org/10.1371/journal.pgen.1008895

50. Lefebvre, L. (2013). Brains, innovations, tools and cultural transmission in birds, non-human primates, and fossil hominins. *Frontiers in Human Neuroscience, 7*. https://doi.org/10.3389/fnhum.2013.00245

CHAPTER 7

Make Your Stash: Resource Guarding

I.

I must say, Whitehaven Beach really delivered on the whiteness and the haven. After a long and rough journey there, we decided to stay at the anchorage at Chalkie's for two nights. This meant we had a whole day to do nothing but frolic in the sun. There wasn't another soul in sight. We had all to ourselves what resembled a Microsoft screen saver—some desolate, sandy isle dropped amid turquoise waters. *Do those places really even exist?* I used to ask myself while staring at the computer monitor. Yes, they do.

On our second morning at Chalkie's, I lay in bed and listened to my husband produce soft snores next to me. The gentle rocking of the boat, reminiscent of a cradle, kept lulling me back to sleep. After half an hour or so of drifting in and out, my morning thirst became stronger than my desire to stay in bed.

I hopped down from the bed and tiptoed to the galley sink. I grabbed a small plastic tumbler from the cabinet and placed it under the tap. The water filled the cup halfway before becoming erratic. It squirted out of the tap in pulses, almost like there were pockets of air in it. I wondered if something was playing up with the tap.

I turned the mixer off, waited a few seconds, and tried it again. Again there were bursts of water rather than a steady stream. Eventually I filled my cup, but the variable pressure meant that I splashed water all over the galley counter. *How annoying*, I thought.

I glanced at the bookshelf on the port side as I put the cup of water to my lips and felt the cool water glide over my tongue. Looking at the red cover of *Bird by Bird* by Anne Lamott on the shelf, I thought maybe I could get some writing done while my husband slept. I put my empty cup down and pulled the book from the shelf. Immediately I felt calmer just having its thin, bendy form in my hand.

I ascended the companionway steps and opened the doors. Just as I did, I heard the door of the forward cabin open behind me. *He's up.* Normally my husband's emergence would mean that my writing window had closed, as morning chatter would soon commence. But on this particular morning, he looked sleepy, not chatty, so maybe I could still squeeze in some writing.

He pulled his toothpaste out of the hanging pocket next to the companionway and moved to the galley sink to brush his teeth. As soon as he turned on the tap, he was greeted by the same erratic water flow that I had experienced earlier. It flattened the glob of toothpaste on his toothbrush and distributed flecks of water all around the sink. He looked at me, puzzled.

"Yeah, I noticed that earlier," I said. "It looks like the water is playing up."

I knew that without being asked, he would go straight into

investigating the problem. (He's a MacGyver—I'm lucky like that.) After he finished brushing his teeth, he unlatched the bottom of the companionway stairs and lifted them. The stairs reminded me of the hydraulic door of a Lamborghini. I grinned, thinking how the boat was way too fancy for us. With the stairs looming over his messy head of hair, he turned the valve that was "on" to the "off" position, and he turned the other one to "on". I remembered switching over the tanks just a couple of days earlier, so I silently wondered whether this would solve the problem.

He went to the sink to test the tap again. This time the spurting was even worse. In fact, after a few seconds, there was nothing but light spit coming out of the tap. We looked at each other, having both realized what had happened. We were out of water.

Running out of fresh water on a boat is a problem. Larger boats that go on long passages often have desalination units on board, which use a process of reverse osmosis to turn salty, impure sea water into fresh drinking water. Those who don't have desalination units need to plan ahead. You calculate how many days you're going to be at sea, and you ration your fresh water accordingly. Every time you turn on the tap to wash your hands, rinse dishes, or take a shower, you're dipping into your fresh water supply. When we first boarded the boat at Abell Point Marina, the guy from the charter company—I think his name was Jason—had warned us to watch our water use. And yet there we were, on the other side of the islands, bone dry.

I opened the galley fridge to see if there were any stores of drinking water left. I'm a compulsive water drinker, and I always get anxious about the possibility that I might not be able to drink my 4 liters a day. Most days I don't drink this amount, but I like to know that I *can*. Inside the galley fridge were our red and blue aluminum water bottles that we brought from home. Both were full of water. Unlike me, my husband had probably not had a sip of his water bottle since we had been there. I dug deeper into the fridge. Down at the bottom there was a nearly frozen 1.5 L plastic bottle that was full. All up, we had 3 L. We would need more.

We decided to lift anchor and find water. We made a plan to motor at full speed towards the marina at Hamilton Island, the only other marina in the Whitsundays apart from Abell Point, so we could refill our water tanks. It took us an hour of motoring before we started to approach the island.

Although the motor was loud, I was startled by an even louder sound as we approached. I looked up. Flying above us at what seemed like touching distance was a Virgin Airlines jet coming in for landing at Hamilton Island. It was the first plane I had seen since air travel was cut less than a week earlier.

Where is this plane coming from? I wondered. Surely they were not bringing more tourists to Hamilton Island. I couldn't think of anything worse than a small island crammed with tourists when it came to preventing the spread of an infectious disease...or really just in general. Maybe they were evacuating people. If so, they must have really wanted them out of there fast.

At this point, it occurred to me to check the news. I went below and disconnected my phone from the charging cable. I typed "Hamilton Island" into Google and hit News.

The news feed that appeared grasped my attention. Hamilton Island was closing, and everyone who was a non-resident was being evacuated. I could understand canceling international travel, but making everyone go home from their holidays within Australia? *Surely no one all the way up here has coronavirus*, I thought.

I plopped down on the sofa and prepared to figure out our next move. I thought to call the Hamilton Island marina just to make sure that we could get in and refill our water. *If there's an empty berth*, I mused wishfully, *we might even be able to stay the night and go out to eat somewhere on the island.*

I searched for the marina's website. At the top of their home page appeared an advisory notice in bold:

>>>As of today, the marina is closed until further notice due to coronavirus.<<<

I turned to my husband, who was playing on his phone on the other sofa, and I showed him the notice. We looked at each other, confused. What in the world was going on? Marinas closing? We came to the Whitsundays to escape, but suddenly it was starting to feel more like we were trapped.

I ruminated on our dire situation. *Here we are, surrounded by water, and yet, it's useless to us. The very water that we need, the only water that we can drink, is inaccessible to us.*

I had many times before mulled over the reality that you can't always get what you want in life, but what happens when you can't even get what you need?

II.

Chile is a country that has been both blessed and cursed with natural resources. As an idealistic college student studying abroad there, I saw capitalism as the root of all poverty. At the time I had every intention of going into medicine and becoming some Che Guevara-type physician to the poor who had no doctor. Had I been more serious about medicine, I would have taken more courses in biochemistry and physiology while in Chile. But instead, I was lured by economics.

My lecturer was Felipe Kast, a Harvard-educated Chilean economist. He had spent his life steeped in economics. He believed better economic policy was the way to a more equal society, one in which the poor would have access to opportunities to move themselves out of poverty. It wasn't that simple, though. As I would come to learn during my semester in Chile, money and power are inseparable. If you try to create economic opportunities for the poor, you're not just playing with other people's money—you're playing with their power. Start messing around with the distribution of a society's resources, I would learn, and all hell breaks loose.

Prior to the military coup of 1973, Chile was considered the exemplar of democracy in Latin America. For most of its history, Chile was governed by elected presidents who all obediently stepped down at the end of their prescribed terms of office and handed over power to their successors in accordance with the constitution. Other democratic governments in Latin America had been toppled by military juntas in the 1960s—Brazil (1964), Argentina (1966), Peru (1968), and Bolivia (1969)—but not Chile. People insisted that Chile was different. Chile had a strong constitution and a tradition of respecting the law.[1] The political pundits in Washington, D.C., assumed they didn't need to worry about Chile.

It wasn't until a socialist doctor, Salvador Allende, won the Chilean presidential election in 1970 that Washington raised an eyebrow. Allende was the grandson of a founder of the Chilean Radical Party. Ever since his time in medical school in the 1930s, Allende had been concerned by the plight of the poor working class. He had read the works of Karl Marx and Leon Trotsky, and upon finishing medical school, he helped to found the Chilean Socialist Party.[2] Once elected to the Chilean Senate in 1945, Allende introduced legislation that established the first program in the Americas to guarantee universal health care.[2,3] He was a socialist to the core.

Allende's socialist rhetoric didn't sit well with U.S. President Richard Nixon. After the Cuban Revolution of 1959, several countries in Latin America had fallen like dominoes to the influence of socialism. These baby socialist countries were being summoned by Soviet fingers, and this worried Nixon. Allende's triumph in Chile's presidential election only added to his anxiety. At first, Nixon considered a coup, but when the U.S. Central Intelligence Agency (CIA) warned that a coup was unlikely to succeed, Nixon asked the CIA to support the opposition to Allende through their network of allies in political parties and the media.[4]

It didn't take much work at all to support the mounting opposition to Allende. Under Allende, the Chilean economy grew weaker and weaker. In an effort to rescue the economy, Allende increased government spending and carried out land reforms. His plans backfired. Business owners fled the country, and landowners let their properties go to shambles, fearing that they would be seized. His policies hurt not only the rich but the poor and the middle class. The Chilean congress begged the military to put a stop to Allende's reforms.[5]

By late 1972, Chileans were experiencing shortages of basic goods, and the threat of a coup became imminent. In one memorable protest known as the March of the Empty Pots and Pans, thousands of Chilean women took to Calle Moneda and marched towards the presidential palace, banging on their empty pots and pans to protest the food shortages. The Chilean truckers' union went on strike in October, and within 2 weeks, bus and taxi drivers, healthcare workers, engineers, and pilots joined the strike. High-ranking military officers and most of Allende's cabinet resigned over the following months. Left alone, Allende was forced to appoint a new commander in chief of the army—a fiery young man named General Augusto Pinochet.[5]

Throughout the night of September 10, 1973, the CIA station in Santiago received several calls with cryptic messages like, "The baby will be delivered tomorrow," and "Uncle Jonas will be in town tomorrow." Finally, at 8:00 a.m. the next morning, the station received a report that an uprising had started among the Chilean navy in Valparaíso, a coastal city just over an hour away from Santiago. Barricades went up around the U.S. embassy. By 9 a.m., naval infantry had reached the streets of downtown Santiago and surrounded Palacio de La Moneda, Chile's presidential palace.[4] Traffic came to a standstill as fires blazed throughout the capital. The coup had begun.

Trapped inside the Palacio de La Moneda, Allende remained hopeful that parts of his military would not abandon him. He tried to phone his newly appointed head of army, General Pinochet, but he couldn't get hold of him. He assumed that Pinochet had been captured. Little did he know, Pinochet was outside orchestrating the coup.[5]

Just before noon, the Chilean Air Force joined the coup by firing rockets into the presidential palace. At 2:00 p.m., the military finally stormed Palacio de La Moneda.[4] But having taken his own life with an AK-47 that Fidel Castro had given him, Allende was already dead.

With Pinochet's assumption of leadership, a new era began in Chile. Everyone in Washington was delighted at the change of power, especially Nixon. For the United States, it symbolized a triumph against socialism—Chile was no longer at risk of being pulled into the Soviet Union's orbit. Nixon expected that after Pinochet had stabilized the economy, elections would be held, and Chile's long history of democracy would resume.[4] But he was wrong.

Soon after gaining power, Pinochet militarized the state and established a Gestapo-like agency, called the Directorate for National Intelligence, to exterminate the Chilean left. In successive waves, Pinochet ordered the military to abduct, disappear, and kill more than one thousand leaders from Allende's party and other left-leaning parties.[5]

Watching the coup and its aftermath unfold on the television were the so-called "Chicago Boys." They were 30 Chilean students who had studied at the University of Chicago under Nobel Prize-winning economist Milton Friedman. Friedman was a proponent of free market economics—the idea that markets are efficient by nature and governments should keep their hands off the economy.[6] After the Chicago Boys finished their studies, many of them returned to Chile to disseminate the philosophy of free market economics. Of course, Allende wouldn't hear a word of the free market school of thought. But after Pinochet's military junta toppled Allende in 1973, the Chicago Boys were suddenly faced with a window of opportunity.

Pinochet wanted a completely new economic model for Chile. He aimed to wash away every stain of the failed socialist experiment that Allende had carried out, and he saw the Chicago Boys as his way to do this. With Pinochet's appointment of Chicago Boy Sergio de Castro as Minister of Economic Affairs in 1975, Chile's neoliberal experiment began.[6]

Miguel Kast, the father of my economics lecturer, was one of the Chicago Boys to spearhead the experiment. As a German-born immigrant, he had lived through extreme poverty when his family first arrived in Chile, and because of this, he became a fervent believer in the application of free market economics to reduce poverty. In 1973, as his studies at the University of Chicago came to a close, Miguel saw the news of the successful coup.[7] He felt he had a duty to return to Chile and implement the free market economic policies that he had learned.

Miguel secured a position at Chile's State Planning Agency, ODEPLAN, and he began working painstakingly to develop policies that would give the poor a ladder out of poverty. When he wasn't at ODEPLAN, Miguel was usually in the chambers of a government or church authority, passionately convincing them of the need to approve certain policies.[7] Nevertheless, Miguel's efforts could not save Chile's crumbling economy.

In a reaction to Allende's socialism, Pinochet had transferred billions of dollars of state assets into the hands of private corporations.[5] The country's natural resources, like copper, which formed the economic backbone of Chile, were sold to transnational corporations virtually tax free.[8] Chilean banks began to take advantage of loopholes in the regulations and offer loans to businesses and people freely, regardless of their ability to repay them. When, in turn, these domestic banks needed money, it was easier than ever to borrow from foreign banks. Between mid-1979 and 1981, the amount of loans relative to the country's gross domestic product (GDP) almost doubled.[9]

As Chile's neoliberal economic situation became graver, Miguel became more contemplative and preoccupied. He started spending work days attending mass and reading poetry. In November 1981, he asked 10 of his closest colleagues to join him daily in El Sagrario church, to pray during the Month of Mary. He asked them to pray for Chile.[7]

Outside the walls of the church, the economy was headed towards a disaster. The loose credit situation was spiraling out of control. From early 1982, the inflation rate in Chile began to rise. Foreign banks began to tighten their lending practices, which meant that Chilean banks no longer had access to foreign loans. With multiple banks going under and the first signs of recession creeping in, Pinochet decided he needed to bring in Miguel Kast to run Chile's Central Bank.[7]

As president of the Central Bank, Miguel took it upon himself to go around convincing everyone, both inside Chile and abroad, to have confidence in the Chilean economy. He implemented a program of economic recovery which involved taking over some of the private banks' riskier loans, but it was too late. The economy was already avalanching. Between 1982 and 1983, Chile's GDP fell by over 20%. By late 1983, unemployment climbed to over 30%, and the percentage of the population living in absolute poverty increased to 55%.[9] Miguel became more and more tired, arriving home late at night only to fall into bed. At the same time, he began to seek more spiritual guidance from the Church.[7]

Finally, Miguel met with Pinochet and formally resigned. Four months later, he was diagnosed with cancer. He was flown to the United States to receive care at Georgetown University Hospital. While there, he received a visit from his former economics lecturer at the University of Chicago, Arnold Harberger. On the day of Arnold's visit, Miguel wrote in his journal, "Today ... came Arnold Harberger. It was a short, but deep [visit]. I told him that Chicago should be 'humanized,' teaching that part to the students."[7] Harberger agreed with Miguel and, crying, told him that he was trying to recreate something like that in his new teaching position at the University of California, Los Angeles.[7]

What Miguel Kast was getting at, just 7 months before his death, was that economies often need humanizing if they are going to function in a way that meets the needs of humans. However, taking a humanistic approach to the economy is hard work, because instead of leaving the market to chance, we are forced to decide: As a society, how should we distribute our resources? Our food, our water, our wealth? Are we going to share these resources, and if so, how? Today, over half of Chile's income is held by just 20% of the population.[10] As Miguel Kast learned from his experiences, even if a country's overall resources are booming, they don't necessarily trickle down to everyone. We need to humanize our economies by making a conscious decision to share our resources in a more humane way.

Both Allende and Pinochet were trying to answer the question of how Chile should share its resources. Allende put the government in charge of the country's resources, believing that this would yield the best outcomes for the poor, while Pinochet let private owners do as they pleased with the resources, which sent the economy out of control. Both were after the same outcome—a better Chile—but neither knew how to manage the country's resources in a way that met the needs of society.

The example of Chile demonstrates that the way in which we decide to control our resources can have brutal consequences on the lives of human beings. A society of otherwise cooperative humans can quickly rip apart at the seams if we begin to put the quest for resource control over the value of human life. A poignant example: In 1991, one year after General Pinochet lost a Chilean plebiscite and finally relinquished his presidency, over 100 human bodies were recovered from a mass grave site now known as Patio 29 in Santiago's General Cemetery. The bodies had been secretly buried there in the 1970s during Pinochet's war against the political left. In some cases, two bodies had been shoved into a single casket. Just after the bodies were discovered, a young television reporter rushed up to Pinochet and asked him to comment on the finding that some of the coffins contained two bodies. Pinochet just chuckled and responded, "How very economical!"[5]

III.

Some of humanity's decisions about economics still send chills down our spines today. But humans aren't the only ones grappling with the difficult issue of resource management. Gray wolves (*Canis lupus*) and domestic dogs (*Canis familiaris*) diverged only recently, sometime in the last 40,000 to 15,000 years,[11] but they now behave very differently when it comes to managing their resources.

When I first met Leo at the breeder, the baby basset hound in my arms felt like a sack of new potatoes. His silky folds of skin and floppy caramel ears spilled over my forearms in every which way. He took slow, calm breaths from a soft, pink belly. He gazed up at me with his sleepy brown eyes, and I knew I had found my new best friend.

Training Leo was far from easy. He'd go to nip my hand with his tiny razor teeth, and I'd yell "No!" at him in the sternest tone I could muster. This was the teaching method I had learned growing up; in those days, almost everyone agreed you needed to "show the dog who is boss." Leo, however, would not be bossed. Standing no taller than my knees, he didn't hesitate to glare up at me and bark back with the same vigor that I had scolded him. I was blown away. A puppy who talks back to you? What was this creature?

I turned to online forums for guidance on how to deal with him. Many people said that basset hounds in particular have a tendency to be "stubborn" and "possessive." If you don't train this "possessive" behavior out of them, people said, they can later develop behavior problems with guarding their food, sometimes aggressively. It can be dangerous, others said, as they can swallow rocks and non-food objects in an effort to guard them. *Great*, I thought. *Now I have to worry about him eating rocks.*

Given his general resistance to my approach, training any behavior in Leo was going to be a challenge. Yelling at him was doing no good, so I figured I'd give whispering a try. I bought *Cesar's Way* by celebrity "dog whisperer" Cesar Millan, and I began reading it one weekend. Millan suggested that dogs, like wolves, are accustomed to power structures or hierarchies in their social groups, where one individual is at the top (the "alpha") and the others are lower in the pecking order. According to Millan, Leo's defiance of my authority could be explained by his belief that he was the alpha of our pack. At this point, I looked over the spine of the book at Leo's pudgy body sprawled across the carpet. How could this little sack of potatoes really think he was my ruler?

Millan insisted that for Leo to become an obedient and calm puppy, I would need to re-establish the proper power dynamic between us, restoring myself as the alpha of our pack. So, for a very ridiculous week,

I made Leo wait before walking through every doorway and sit like a statue before being allowed to eat. I walked around the neighborhood like a burly man, with little Leo by my side, trying to take large, audible strides so that Leo would learn that I was a dominant force. He seemed confused, if not amused, by my change in behavior. One week in and he was no more obedient than when I'd first brought him home.

After my initial frustration passed, I found the whole thing laughable too. I couldn't fool this animal into thinking I was his male wolf leader. It was pretty clear that I was the same person who brought him home on my lap just a few weeks earlier and who doted on him around the clock (when he wasn't being a little monster). I began to suspect that the dominance approach didn't have much scientific merit, so I decided to sift through the research on wolves and dogs for myself.

Studies of wolves have historically been carried out in captive settings, with researchers penning wolves with other unrelated wolves.[12, 13] In these tense, artificial settings, wolves often displayed what looked like dominance when a valued resource like meat was presented, with some individuals always feeding first and not allowing others to access it. This pattern became even more pronounced during times of food shortage, leading researchers to conclude that wolves form hierarchies of dominance within their packs over access to food.[13] But canine biologist David Mech, who had been observing wolf packs in the wild for most of his life,[14] was unconvinced by these findings of so-called dominance coming out of captive settings.

To better understand the social dynamics of wolves, Mech devoted several decades to watching how wolves behave in the wild. In the 1960s, he began using radio collars to track the movements of individuals and packs of wolves on Isle Royale, Lake Superior. Later he used the same method to study the reintroduced wolves in Yellowstone National Park. During these studies, Mech gained some of the first insights into how wolves share resources when they're living in groups of their own choosing.[14]

Across his decades of observations, Mech found that rather than competing for resources by jostling their way to the top of a pecking order, wild wolves often cooperate.[15] In the wild, most wolves live in families, also called packs, which are comprised of a mated pair with all of their offspring.[15] That being said, just as with humans, there are many kinds of wolf families, including ones with only one parent and ones who have adopted non-relatives.[16] Mech found that within wolf families, contests over power or dominance are rare.[17] Living in tight-knit groups, the wolves whom Mech observed cooperated for many aspects of their lives,[15] including hunting,[18] rearing pups,[19] and defending their territories.[20]

To understand why wolves cooperate, it's helpful to look at the resources they value most and how cooperation helps to gain or prevent the loss of those resources. Food is a valuable resource for all animals, but it is especially valuable when it's scarce. Wolves feed on some of the largest hoofed mammals, including bison, elk, deer, caribou, musk oxen, and moose. They hunt in groups to take down these strong animals, but even so, most hunts are unsuccessful.[15] Accordingly, wolves in the wild live a feast-or-famine existence. Mech found that 7–65% of wolves he observed died each year, in large part due to starvation.[21] Because starvation is such a great risk, wolves must cooperate to get whatever food they can to survive.

Apart from being unpredictable in terms of success, hunting is also a dangerous activity. To prevent harm to themselves, wolves need to coordinate their actions seamlessly during the hunt.[22] They pay close attention to each other, following each other's gaze and imitating each other's actions.[23] Hunting is a cooperative skill that wolf pups learn from 4 to 10 months of age, when they begin accompanying their parents on hunting excursions.[16]

Mech observed that, after a successful hunt, wild wolves share their food by distributing the meat among pack members and regurgitating it to the pups. In general, the larger the prey, the more wolves it can support, so the larger the group size tends to be.[15] By cooperating to acquire and share their food, wolves are able to give themselves a better chance of survival in an otherwise unpredictable environment.

For wolves living on the Eurasian continent during the Upper Paleolithic, resources became much more predictable with the arrival of modern humans, *Homo sapiens*. Although scientists now suspect that *Homo sapiens* arrived to Eurasia as early as 210,000 years ago,[24] evidence of humans interacting with and domesticating wolves only appears in the last 40,000 to 15,000 years.[11] The predominant theory of how dogs were domesticated, which is based on evidence of paleolithic dog remains and studies of wolf behavior, is that humans living in Eurasia repeatedly collected wolf pups from their dens and fed them into adulthood, thereby socializing them to humans.[25]

Stealing wolf pups is theoretically quite doable, because from about 3 weeks of age, wolf pups are left alone in their dens for varying periods as their mothers go out to hunt. Between 3 and 5 weeks of age, wolf pups become less reliant on their mothers' milk and start to explore outside their natal dens. During the life stage from 3 to 11 weeks, referred to as the "critical period of socialization," wolf pups begin to form attachments to their various family members, and they learn how to cooperate with each other through social play.[16] By collecting wolf pups around this critical life stage and providing them with nurture and social interaction, humans could have shaped the wolves' social tendencies and directed them towards humans. In return, these domesticated wolves might have helped human communities to hunt and transport large animals. The domesticated wolves would have also been useful for guarding humans and their settlements from predators, alerting communities to the presence of intruders, cleaning up settlement debris, and simply providing companionship.[25]

The wolf pups' captors presumably kept and raised only those pups who had the most desirable traits. Through human eyes, this usually looks like docile temperament, greater hunting abilities, or cuteness, all of which are influenced by genetics. In addition, early humans likely hand fed the pups they chose to keep in response to their showing desirable behaviors, which made the pups more likely to repeat those behaviors as they got older. This highly effective learning strategy, known as positive reinforcement,[26] is now recognized as best practice in the animal training community. It's funny to think that our ancestors were probably doing it 40,000 years ago. Studies of teeth from paleolithic dogs found in the Czech Republic confirm that the dogs ate more brittle foods than wolves of their time did. This suggests that the dogs were regularly fed bones and scraps by humans, while wolves in the area consumed mostly mammoth meat.[11] Without consistent feeding by humans, it's likely that any wolf pups tamed by humans would have eventually wandered into the wild to breed, never to return. On the other hand, well-fed pups would have returned to humans as a reliably good food source.[25] And so the interdependence of man and dog began.

From the time of their divergence with wolves, dogs have had access to an extremely predictable resource—human beings. Dogs know when and where they can expect to be fed each day. Meanwhile, wolves have continued to face the unpredictability inherent in hunting large game, not knowing if or when they'll get to eat. Is it any wonder, then, that modern dogs are less prone to cooperate with each other than wolves are?

You can see the difference if you place wolves and dogs in a loose-string experiment, where two animals need to pull the two ends of a rope to move a tray and gain access to food. Wolves are highly successful in cooperating with each other to get the food. Dogs, however, perform rather poorly in this task, suggesting that dogs are less cooperative with members of their own species.[27] Unsurprisingly, when pet dogs were trained to do the task with their owner instead of another dog, they performed much better.[28]

Dogs are not only less cooperative with other dogs in acquiring food; they are also less likely to share food with each other once they have it. When dogs and wolves were tested in pair-wise food competition tests, one dog of each pair monopolized the food resource, feeding alone most of the time, whereas wolves fed together for longer.[23] In a similar study, individual dogs were found to guard food for themselves, while wolves allowed each member of the pack a chance to eat.[29]

The different degrees of cooperation in wolves and dogs may have something to do with the predictability of food in their environments.[29] Wolves survive on hunting large prey, which are sparsely distributed and appear unpredictably. As such, cooperation in hunting and food sharing are essential parts of their survival.[19] Dogs, on the other hand, rely almost entirely on sources provided by humans.[29] Dogs' food is usually concentrated in location (near humans) and is highly predictable (appearing every day of the year). Therefore, dogs have little need to cooperate with other dogs or to share their food.

Looking back at the history of the human-dog relationship, we can see that, rather than one party dominating the other, dogs and humans were (and still are) essentially business partners. We exchange resources for our mutual benefit. Dogs give us their protection, entertainment, and loyalty, and we give them food, shelter, and care. What I began to see from my reading is that for both parties to feel happy and secure, this resource exchange needs to be predictable.

Of course, I thought. *Predictability.* If there is a way for animals to make our lives more consistent, our food more reliable, and our survival better guaranteed, then of course we're going to go for it. For animals, predictability is closely tied to our expectations and our perception of control, so altering predictability can have major impacts on our wellbeing.[30] After this revelation, I began to apply positive reinforcement and predictable signaling to my exchanges with little Leo. Soon he came to learn what he could and could not expect of me, and I, of him. Our mutual understanding deepened, and our business transactions became considerably smoother.

The research I uncovered on dogs and wolves got me thinking beyond just Leo, though. If over the last 40,000 years, dogs have changed their social behavior around sharing resources, becoming less cooperative with each other, then is it possible that modern humans have undergone such a change too?

IV.

Like dogs, humans crave predictable resources. And yet our own species, *Homo sapiens,* was born out of unpredictability.

Anatomically modern humans did not emerge at just one moment or place. Archaeological research has found that *Homo sapiens* was constructed gradually and in a piecemeal fashion across Africa. There was no "great leap forward," no single arrival of features that made us distinct from and superior to all primates who had come before. Instead, our small, flat face and pointed chin arrived at a different time to our rounded skull, which arrived at a different time to our narrow trunk and pelvis.[31, 32] Modern humans coalesced slowly, like a jigsaw puzzle on a side table that gets completed in random, whimsical bursts.

The complex birth of our species makes sense if we consider the mosaic of climates into which *Homo sapiens* were born. Sediment records from East Africa indicate that between 400,000 and 320,000 years ago, ecological resources both on land and in the water began to fluctuate wildly.[33] Variability in rainfall caused lake levels to change frequently and landscapes to shift back and forth between woody vegetation and grasslands.[33] Habitable zones appeared and disappeared. From this time, dominant megagrazers, some weighing over 900 kg, began to disappear from East Africa. Meanwhile, the smaller, mixed-grazing herbivores who were more resilient to drought proliferated.[34] Suddenly, hunting large animals was no longer a predictable way of life for African hominins. It was during this window of intense unpredictability that our species is thought to have emerged.

The oldest remains of *Homo sapiens* have been uncovered at Jebel Irhoud, Morocco, dating back to 349,000–281,000 years ago.[35] These fossils represent an early iteration of our species, one in which our flatter, less-projecting face was already present, but the high, rounded braincase typical of modern humans had not yet evolved.[32] Similar in appearance is a skull recovered from East Turkana, Kenya, which has been dated to around 270,000 years ago.[36] More recent *Homo sapiens* fossils, from 230,000 to 150,000 years ago, have been found in the Omo Valley and Herto, Ethiopia, and by this time, humans already had high, rounded braincases.[37, 38, 39]

Behavioral flexibility might have helped *Homo sapiens* to survive the unpredictable climate around 300,000 years ago. During this period of resource overhaul, humans at multiple locations across Africa abandoned Acheulean stone-flaking technology—the longest enduring Paleolithic technology—and replaced it with Middle Stone Age toolmaking technology.[33] This new way of toolmaking involved making small points, blades, retouched flakes, and bone tools. Middle Stone Age toolmaking required better planning and the procurement of rocks from distant sources.[34] Small stone points with retouched bases suggest that our ancestors also began making projectile armaments around this time, which could reflect their need to hunt smaller prey following the loss of large mammals from Africa.[33]

In addition to refining their technology around 300,000 years ago, our ancestors began to expand their geographical range. The composition of Middle Stone Age tools in the Olorgesailie basin reveals that our ancestors started to transport obsidian from multiple volcanic outcrops over distances of 25–95 km.[33] Such long-distance transfer of obsidian indicates that there could have been some form of resource exchange among interconnected social groups across large parts of Africa.[33, 40] It could be that the capacity of *Homo sapiens* to build and maintain long-distance cooperative social networks helped our species to survive resource landscapes that were becoming less predictable in time and space.[34]

By 300,000 years ago, many groups of *Homo erectus* had already migrated out of Africa and expanded throughout Eurasia, but *Homo sapiens* were still confined to Africa.[41, 42, 43] Remnants of the first members of our species to venture out of Africa have been found inside caves in present-day Israel and are thought to be around 100,000 years old.[44] Humans are thought to have moved into this region, called the Levant, during a particularly humid period between ice ages, when major deserts in northern Africa and the Arabian peninsula became less arid and more permeable to humans and other animals.[45] Many archaeologists assert that these early human migrants to the Levant eventually died out, but other humans who migrated to southern Arabia around the same time survived, spreading across India, Southeast Asia, Australia, and Melanesia by about 50,000 years ago.[44, 46, 47, 48, 49] Other dispersals out of Africa likely followed, with humans reaching northern Eurasia by around 45,000 years ago.[50] During their migrations, humans likely interacted with other hominin species, such as the Neandertals and Denisovans.[50]

When humans first moved into northern Eurasia, the conditions were relatively favorable for them. But over the following millennia, the Earth gradually slipped into an ice age. The coldest, driest period happened some 20,000–15,000 years ago and is known as the Last Glacial Maximum.[51] During the Last Glacial Maximum, glaciers covered 25% of the Earth's land area. These glaciers tied up huge volumes of sea water, which caused sea levels to fall 120 m lower than they are today.[52] While lower sea levels helped our ancestors to move around the globe across land bridges, it also meant there was less ocean to mediate temperatures and rainfall. During this time, many of the game species on which humans depended began to die out. Similarly, the human population is thought to have shrunk in many places, especially in the northern latitudes.[53]

Different locations experienced the Last Glacial Maximum differently due to their elevation, topography, and proximity to lakes. Even in the northern regions of Europe and Asia, there were probably favorable pockets where humans could survive.[53] These pockets would have been few and far between, however, as few locations on Earth would have been lush enough and with consistent enough resources for our ancestors to settle for long.

During this period of adversity, again the behavioral flexibility of humans came to their aid. As their environment became colder and the large animals on which they depended became less abundant, some groups of humans began to cache or stockpile food in order to insure themselves against times of scarcity. It has often been assumed that our habit of accumulating material resources began with the advent of agriculture, but archaeological evidence suggests that humans began stockpiling much earlier, in the lead-up to the last ice age. A 25,000-year-old circular mammoth-bone structure discovered at Kostenki 11 in Russia, which is surrounded by at least three large pits containing mammoth bones, indicates that humans began accumulating and storing food in times when resources were most abundant.[54] In some places, food caching might have been helped by fermentation, as was found at an early human settlement in southern Sweden.[55]

Why stockpile resources? Archaeologist Lewis Binford theorized that in colder environments, where resources are clumped in time and space, it makes sense for societies to do massive amounts of hunting during peak seasons, accumulate a surplus, and store the food for the rest of the year. Binford reasoned that this level of resource caching is not needed in warmer climates, where nomadic lifestyles are more effective.[56] He partially based this theory on the settlement systems of 168 modern hunter-gatherer groups. For example, the Alaskan Nunamiut, with whom Binford lived for several years, organize their lives around the seasonal migrations of caribou. The Nunamiut will send out a group of young men and women for 5 days to hunt and process caribou, and then they'll return to their home base with the food to share and store.[57] The storage of food at a home base allows these groups to accumulate a surplus, which helps them to weather times when resources are less abundant.

The more you stockpile a critical resource like food, the more important it becomes to defend it. The theory of economic defendability, which was put forth by ethologist Jerram Brown in 1964 in relation to birds, posits that when animals have a resource that is densely concentrated in the environment and is predictable, they are more likely to become territorial about it and to engage in active defense of their territories.[58] It's difficult to guard a resource if you're alone. At some stage you'll need to leave the resource to go and collect more food. Add just one more person, though, and one person can stay and guard the property while the other goes out and forages.[59] Zooarchaeologist Curtis Marean postulates that the dependence on clustered resources might have fostered even greater cooperation among unrelated humans, as they had to defend their territories and engage in conflict against other groups to gain and protect their resources.[57]

Around 15,000 years ago, our ancestors finally got a reprieve. The periodic bending of Earth's orbit brought Earth slightly closer to the Sun than it had been before, and the increased sunlight caused the glaciers to melt. Because water does not reflect as much sunlight as ice and absorbs heat because of its dark color, the additional water acted as a moderating force for temperature and humidity, making our ancestors' climate more stable. From about 11,700 years ago, sea levels rose, lakes and rivers filled, and our planet entered a warmer and more stable epoch known as the Holocene.[60] In some places, such as Greenland, this drastic switch from the unstable ice age climate to the stable climate of the Holocene is thought to have happened in the span of only about 40 years.[61] Within a single lifespan, the world of many humans shifted from a chaotic one to an oddly stable one.

Archaeological records show that during this transition from the Last Glacial Maximum to the early Holocene, humans covered substantial ground in their expansion throughout Eurasia.[62] We see evidence of genetic intermixing and resource exchange among groups of humans during the early Holocene. By this time, the practice of building semi-permanent houses and storing surpluses was already well established, so the interaction among groups of humans was likely very different to our early hunter-gatherer days. With the favorable climates provided by the Holocene some 13,000–10,000 years ago, the cultivation of grains and vegetables on a predictable basis and keeping of farm animals became possible at several locations around the world within a relatively short period. Suddenly, resources became abundant and steady for humans.[63, 64] Despite this new abundance of resources, the practice of storing resources "just in case" had likely become cemented in many human cultures, a relic from the resource uncertainty of the last ice age. Humans continued to amass resources.

We have seen in dogs how a change in resource predictability can have a major impact on the cooperative behavior of species. Like wolves, humans evolved in a highly unpredictable environment for millions of years, and we needed to cooperate to access food and survive such unpredictability. But if suddenly, within the last 12,000 years of the Holocene, the resources on which our ancestors depended became strangely predictable—concentrated in space and time—then maybe we no longer needed to cooperate from an early age like wolf pups do (or so we thought). Maybe, like dogs, we began to exert greater control over the resources in our vicinity and to guard them more closely.

The first clear evidence of resource guarding is the oldest known walled city of Jericho, which is some 8,000 years old. This wall represents the first communal effort to fortify a settlement. Fortification against what is still unknown—some have suggested that it was defense against neighboring populations in the Jordan Valley, whereas others insist that the wall was more likely built to protect the city against floods.[65] In addition to the wall, there was an 8.5 m stone tower at the center of Jericho, around which storage facilities were built. The locations of the tower and storage facilities suggest that the resources were publicly owned or at the service of the community, but we don't know for sure.[65] Dense and predictable resources kept in storage facilities provide communities with better nutrition year round, allowing women to bear more offspring and more of their offspring to survive. This can lead to population growth, which is handy because more people makes it easier to defend the village, especially if they need to defend against intruders or go to war with neighboring villages.[57] However, there is no evidence for war in Jericho around 8,000 years ago; rather, features of their art, rituals, mortuary practices, and prestige goods indicate that humans in the Near East and the Levant maintained peaceful relationships over long distances through gift giving, exchange, and trade.[65]

Apart from establishing the practice of accumulating and guarding resources—a practice which our species has since taken to destructive extremes—the transition from the Last Glacial Maximum to the Holocene could have had other effects on the human psyche. In much of the animal kingdom, our expectations about predictability are closely tied to our sense of control. If, starting in the Holocene, humans began to perceive their world as more predictable, and importantly, if they had already established the practice of caching resources, then perhaps they started to feel a greater sense of control over their environment. Focused just on their little slice of time, they might have begun to think, as the fathers of the scientific revolution did, that the world is a rather orderly and predictable place. Under this **Illusion of Security**, it is easy for humans to forget the chaotic dynamics of nature, which are mostly unpredictable. Concentrating only on the stability of present conditions, like a dog living in the predictability of his owner's care, early Holocene humans might have begun to forget the utter chaos from which our species was born.

CHAPTER 7 REFERENCES

Free Resources

Mech, L. D. (2008). *"Alpha" wolf?* Eduweb. https://www.youtube.com/watch?v=tNtFgdwTsbU

Further Reading

Bellwood, P. (2022). *First farmers: The origins of agricultural societies* (2nd ed.). Wiley-Blackwell. https://www.wiley.com/en-au/First+Farmers%3A+The+Origins+of+Agricultural+Societies%2C+2nd+Edition-p-9781119706373

Eaton, B. (2011). *Dominance in dogs: Fact or fiction?* Dogwise Publishing. https://www.dogwise.com/dominance-in-dogs-fact-or-fiction/

Graeber, D., & Wengrow, D. (2021). *The dawn of everything: A new history of humanity.* Farrar, Straus and Giroux. https://us.macmillan.com/books/9780374157357/thedawnofeverything

Mech, L. D., & Boitani, L. (Eds.). (2003). *Wolves: Behavior, ecology, and conservation.* University of Chicago Press. https://press.uchicago.edu/ucp/books/book/chicago/W/bo3641392.html

Pelc, K. (2022). *Beyond self-interest: Why the market rewards those who reject it.* Bloomsbury Publishing. https://www.bloomsbury.com/au/beyond-selfinterest-9781526648143/

References

1. Hilbink, L. (2003). An exception to Chilean exceptionalism?: The historical role of Chile's judiciary. In S. Eckstein & T. Wickham-Crowley (Eds.), *What justice? Whose justice?: Fighting for fairness in Latin America.* University of California Press. https://www.ucpress.edu/book/9780520237452/what-justice-whose-justice

2. Tedeschi, S. K., Brown, T. M., & Fee, E. (2003). Salvador Allende: Physician, socialist, populist, and president. *American Journal of Public Health, 93*(12), 2014–2015. https://doi.org/10.2105/AJPH.93.12.2014

3. Waitzkin, H., Iriart, C., Estrada, A., & Lamadrid, S. (2001). Social medicine then and now: Lessons from Latin America. *American Journal of Public Health, 91*(10), 1592–1601. https://doi.org/10.2105/AJPH.91.10.1592

4. Devine, J. (2014). What really happened in Chile: The CIA, the coup against Allende, and the rise of Pinochet. *Foreign Affairs, 93*(4), 26–35. http://www.jstor.org/stable/24483554

5. Ensalaco, M. (2000). *Chile under Pinochet: Recovering the truth.* University of Pennsylvania Press. https://www.jstor.org/stable/j.ctt3fhg49

6. Silva, P. (1991). Technocrats and politics in Chile: From the Chicago Boys to the CIEPLAN monks. *Journal of Latin American Studies, 23*(2), 385–410. https://doi.org/10.1017/S0022216X00014048

7. Lavín, J. (1986). *Miguel Kast: Pasión* de vivir. Zig-Zag. https://books.google.com.au/books/about/Miguel_Kast.html?id=XL0TAQAAIAAJ&redir_esc=y

8. Cypher, J. M. (2004). Pinochet meets Polanyi? The curious case of the Chilean embrace of "free" market economics. *Journal of Economic Issues, 38*(2), 527–535. https://doi.org/10.1080/00213624.2004.11506714

9. Barandiarán, E., & Hernández, L. (1999). *Origins and resolution of a banking crisis: Chile 1982-86.* https://citeseerx.ist.psu.edu/document?repid=rep1&type=pdf&doi=e1f294bdefa443cbe69c8870693a6086a8be4d03

10. The World Bank. (2020). *Income share held by highest 20% - Chile.* https://data.worldbank.org/indicator/SI.DST.05TH.20?locations=CL

11. Prassack, K. A., DuBois, J., Lázničková-Galetová, M., Germonpré, M., & Ungar, P. S. (2020). Dental microwear as a behavioral proxy for distinguishing between canids at the Upper Paleolithic (Gravettian) site of Předmostí, Czech Republic. *Journal of Archaeological Science, 115*, 105092. https://doi.org/10.1016/j.jas.2020.105092

12. Bradshaw, J. W. S., Blackwell, E. J., & Casey, R. A. (2009). Dominance in domestic dogs—useful construct or bad habit? *Journal of Veterinary Behavior, 4*(3), 135–144. https://doi.org/10.1016/j.jveb.2008.08.004

13. Zimen, E. (1976). On the regulation of pack size in wolves. *Zeitschrift* Für Tierpsychologie, *40*(3), 300–341. https://doi.org/10.1111/j.1439-0310.1976.tb00939.x

14. Mech, L. D. (2002). *The wolves of Isle Royale.* University Press of the Pacific. https://isleroyalewolf.org/wolfhome/books_film.html

15. Mech, L. D., & Boitani, L. (2003). Wolf social ecology. In L. D. Mech & L. Boitani (Eds.), *Wolves: Behavior, ecology, and conservation.* University of Chicago Press. https://press.uchicago.edu/ucp/books/book/chicago/W/bo3641392.html

16. Packard, J. M. (2003). Wolf behavior: Reproductive, social, intelligent. In L. D. Mech & L. Boitani (Eds.), *Wolves: Behavior, ecology, and conservation.* University of Chicago Press. https://press.uchicago.edu/ucp/books/book/chicago/W/bo3641392.html

17. Mech, L. D. (1999). Alpha status, dominance, and division of labor in wolf packs. *Canadian Journal of Zoology, 77*(8), 1196–1203. https://doi.org/10.1139/z99-099

18. MacNulty, D. R., Smith, D. W., Mech, L. D., Vucetich, J. A., & Packer, C. (2012). Nonlinear effects of group size on the success of wolves hunting elk. *Behavioral Ecology, 23*(1), 75–82. https://doi.org/10.1093/beheco/arr159

19. Mech, L. D., Wolf, P. C., & Packard, J. M. (1999). Regurgitative food transfer among wild wolves. *Canadian Journal of Zoology, 77*(8), 1192–1195. https://doi.org/10.1139/z99-097

20. Cassidy, K. A., MacNulty, D. R., Stahler, D. R., Smith, D. W., & Mech, L. D. (2015). Group composition effects on aggressive interpack interactions of gray wolves in Yellowstone National Park. *Behavioral Ecology, 26*(5), 1352–1360. https://doi.org/10.1093/beheco/arv081

21. Mech, L. D. (1977). Productivity, mortality, and population trends of wolves in northeastern Minnesota. *Journal of Mammalogy, 58*(4), 559–574. https://doi.org/10.2307/1380004

22. Mech, L. D., & Peterson, R. O. (2003). Wolf-prey relations. In L. D. Mech & L. Boitani (Eds.), *Wolves: Behavior, ecology, and conservation.* University of Chicago Press. https://press.uchicago.edu/ucp/books/book/chicago/W/bo3641392.html

23. Range, F., & Virányi, Z. (2015). Tracking the evolutionary origins of dog-human cooperation: the "Canine Cooperation Hypothesis." *Frontiers in Psychology, 5*. https://doi.org/10.3389/fpsyg.2014.01582

24. Harvati, K., Röding, C., Bosman, A. M., Karakostis, F. A., Grün, R., Stringer, C., Karkanas, P., Thompson, N. C., Koutoulidis, V., Moulopoulos, L. A., Gorgoulis, V. G., & Kouloukoussa, M. (2019). Apidima Cave fossils provide earliest evidence of *Homo sapiens* in Eurasia. *Nature, 571*(7766), 500–504. https://doi.org/10.1038/s41586-019-1376-z

25. Mech, L. D., & Janssens, L. A. A. (2022). An assessment of current wolf *Canis lupus* domestication hypotheses based on wolf ecology and behaviour. *Mammal Review, 52*(2), 304–314. https://doi.org/10.1111/mam.12273

26. Skinner, B. F. (1938). *The behavior of organisms: An experimental analysis*. Appleton-Century. https://psycnet.apa.org/record/1939-00056-000

27. Marshall-Pescini, S., Schwarz, J. F. L., Kostelnik, I., Virányi, Z., & Range, F. (2017). Importance of a species' socioecology: Wolves outperform dogs in a conspecific cooperation task. *Proceedings of the National Academy of Sciences, 114*(44), 11793–11798. https://doi.org/10.1073/pnas.1709027114

28. Ostojić, L., & Clayton, N. S. (2014). Behavioural coordination of dogs in a cooperative problem-solving task with a conspecific and a human partner. *Animal Cognition, 17*(2), 445–459. https://doi.org/10.1007/s10071-013-0676-1

29. Dale, R., Range, F., Stott, L., Kotrschal, K., & Marshall-Pescini, S. (2017). The influence of social relationship on food tolerance in wolves and dogs. *Behavioral Ecology and Sociobiology, 71*(7), 107. https://doi.org/10.1007/s00265-017-2339-8

30. Bassett, L., & Buchanan-Smith, H. M. (2007). Effects of predictability on the welfare of captive animals. *Applied Animal Behaviour Science, 102*(3–4), 223–245. https://doi.org/10.1016/j.applanim.2006.05.029

31. Andirkó, A., Moriano, J., Vitriolo, A., Kuhlwilm, M., Testa, G., & Boeckx, C. (2021). Fine-grained temporal mapping of derived high-frequency variants supports the mosaic nature of the evolution of *Homo sapiens*. *BioRxiv*. https://doi.org/10.1101/2021.01.22.427608

32. Neubauer, S., Hublin, J.-J., & Gunz, P. (2018). The evolution of modern human brain shape. *Science Advances, 4*(1). https://doi.org/10.1126/sciadv.aao5961

33. Potts, R., Dommain, R., Moerman, J. W., Behrensmeyer, A. K., Deino, A. L., Riedl, S., Beverly, E. J., Brown, E. T., Deocampo, D., Kinyanjui, R., Lupien, R., Owen, R. B., Rabideaux, N., Russell, J. M., Stockhecke, M., deMenocal, P., Faith, J. T., Garcin, Y., Noren, A., ... Uno, K. (2020). Increased ecological resource variability during a critical transition in hominin evolution. *Science Advances, 6*(43). https://doi.org/10.1126/sciadv.abc8975

34. Potts, R., Behrensmeyer, A. K., Faith, J. T., Tryon, C. A., Brooks, A. S., Yellen, J. E., Deino, A. L., Kinyanjui, R., Clark, J. B., Haradon, C. M., Levin, N. E., Meijer, H. J. M., Veatch, E. G., Owen, R. B., & Renaut, R. W. (2018). Environmental dynamics during the onset of the Middle Stone Age in eastern Africa. *Science, 360*(6384), 86–90. https://doi.org/10.1126/science.aao2200

35. Richter, D., Grün, R., Joannes-Boyau, R., Steele, T. E., Amani, F., Rué, M., Fernandes, P., Raynal, J.-P., Geraads, D., Ben-Ncer, A., Hublin, J.-J., & McPherron, S. P. (2017). The age of the hominin fossils from Jebel Irhoud, Morocco, and the origins of the Middle Stone Age. *Nature, 546*(7657), 293–296. https://doi.org/10.1038/nature22335

36. Bräuer, G., Yokoyama, Y., Falguères, C., & Mbua, E. (1997). Modern human origins backdated. *Nature, 386*(6623), 337–338. https://doi.org/10.1038/386337a0

37. McDougall, I., Brown, F. H., & Fleagle, J. G. (2005). Stratigraphic placement and age of modern humans from Kibish, Ethiopia. *Nature, 433*(7027), 733–736. https://doi.org/10.1038/nature03258

38. Vidal, C. M., Lane, C. S., Asrat, A., Barfod, D. N., Mark, D. F., Tomlinson, E. L., Tadesse, A. Z., Yirgu, G., Deino, A., Hutchison, W., Mounier, A., & Oppenheimer, C. (2022). Age of the oldest known *Homo sapiens* from eastern Africa. *Nature, 601*(7894), 579–583. https://doi.org/10.1038/s41586-021-04275-8

39. White, T. D., Asfaw, B., DeGusta, D., Gilbert, H., Richards, G. D., Suwa, G., & Clark Howell, F. (2003). Pleistocene *Homo sapiens* from Middle Awash, Ethiopia. *Nature, 423*(6941), 742–747. https://doi.org/10.1038/nature01669

40. Scerri, E. M. L., Chikhi, L., & Thomas, M. G. (2019). Beyond multiregional and simple out-of-Africa models of human evolution. *Nature Ecology & Evolution, 3*(10), 1370–1372. https://doi.org/10.1038/s41559-019-0992-1

41. Potts, R., & Teague, R. (2010). Behavioral and environmental background to 'Out-of-Africa I' and the arrival of *Homo erectus* in East Asia. In J. G. Fleagle, J. J. Shea, F. E. Grine, A. L. Baden, & R. E. Leakey (Eds.), *Out of Africa I: The first hominin colonization of Eurasia* (pp. 67–85). Springer. https://doi.org/10.1007/978-90-481-9036-2_5

42. Rizal, Y., Westaway, K. E., Zaim, Y., van den Bergh, G. D., Bettis, E. A., Morwood, M. J., Huffman, O. F., Grün, R., Joannes-Boyau, R., Bailey, R. M., Sidarto, Westaway, M. C., Kurniawan, I., Moore, M. W., Storey, M., Aziz, F., Suminto, Zhao, J., Aswan, ... Ciochon, R. L. (2020). Last appearance of *Homo erectus* at Ngandong, Java, 117,000–108,000 years ago. *Nature, 577*(7790), 381–385. https://doi.org/10.1038/s41586-019-1863-2

43. Holloway, R. L. (1981). Volumetric and asymmetry determinations on recent hominid endocasts: Spy I and II, Djebel Ihroud I, and the salè *Homo erectus* specimens, with some notes on neandertal brain size. *American Journal of Physical Anthropology, 55*(3), 385–393. https://doi.org/10.1002/ajpa.1330550312

44. Rabett, R. J. (2018). The success of failed *Homo sapiens* dispersals out of Africa and into Asia. *Nature Ecology & Evolution, 2*(2), 212–219. https://doi.org/10.1038/s41559-017-0436-8

45. Ekshtain, R., & Tryon, C. A. (2019). Lithic raw material acquisition and use by early *Homo sapiens* at Skhul, Israel. *Journal of Human Evolution, 127*, 149–170. https://doi.org/10.1016/j.jhevol.2018.10.005

46. Rosenberg, T. M., Preusser, F., Fleitmann, D., Schwalb, A., Penkman, K., Schmid, T. W., Al-Shanti, M. A., Kadi, K., & Matter, A. (2011). Humid periods in southern Arabia: Windows of opportunity for modern human dispersal. *Geology, 39*(12), 1115–1118. https://doi.org/10.1130/G32281.1

47. Petraglia, M. D., Haslam, M., Fuller, D. Q., Boivin, N., & Clarkson, C. (2010). Out of Africa: New hypotheses and evidence for the dispersal of *Homo sapiens* along the Indian Ocean rim. *Annals of Human Biology, 37*(3), 288–311. https://doi.org/10.3109/03014461003639249

48. Reyes-Centeno, H., Ghirotto, S., Détroit, F., Grimaud-Hervé, D., Barbujani, G., & Harvati, K. (2014). Genomic and cranial phenotype data support multiple modern human dispersals from Africa and a southern route into Asia. *Proceedings of the National Academy of Sciences, 111*(20), 7248–7253. https://doi.org/10.1073/pnas.1323666111

49. Groucutt, H. S., Petraglia, M. D., Bailey, G., Scerri, E. M. L., Parton, A., Clark-Balzan, L., Jennings, R. P., Lewis, L., Blinkhorn, J., Drake, N. A., Breeze, P. S., Inglis, R. H., Devès, M. H., Meredith-Williams, M., Boivin, N., Thomas, M. G., & Scally, A. (2015). Rethinking the dispersal of *Homo sapiens* out of Africa. *Evolutionary Anthropology: Issues, News, and Reviews, 24*(4), 149–164. https://doi.org/10.1002/evan.21455

50. Li, F., Petraglia, M., Roberts, P., & Gao, X. (2020). The northern dispersal of early modern humans in eastern Eurasia. *Science Bulletin, 65*(20), 1699–1701. https://doi.org/10.1016/j.scib.2020.06.026

51. Hubberten, H. W., Andreev, A., Astakhov, V. I., Demidov, I., Dowdeswell, J. A., Henriksen, M., Hjort, C., Houmark-Nielsen, M., Jakobsson, M., Kuzmina, S., Larsen, E., Lunkka, J. P., Lyså, A., Mangerud, J., Möller, P., Saarnisto, M., Schirrmeister, L., Sher, A. v, Siegert, C., ... Svendsen, J. I. (2004). The periglacial climate and environment in northern Eurasia during the Last Glaciation. *Quaternary Science Reviews, 23*(11–13), 1333–1357. https://doi.org/10.1016/j.quascirev.2003.12.012

52. U.S. Geological Survey. (n.d.). *How does present glacier extent and sea level compare to the extent of glaciers and global sea level during the Last Glacial Maximum (LGM)?* Climate. Retrieved December 6, 2023, from https://www.usgs.gov/faqs/how-does-present-glacier-extent-and-sea-level-compare-extent-glaciers-and-global-sea-level?qt-news_science_products=0#qt-news_science_products

53. Buvit, I., Rasic, J. T., & Izuho, M. (2022). Archaeological evidence shows widespread human depopulation of Last Glacial Maximum Northeast Asia. *Archaeological and Anthropological Sciences, 14*(7), 125. https://doi.org/10.1007/s12520-022-01596-1

54. Pryor, A. J. E., Beresford-Jones, D. G., Dudin, A. E., Ikonnikova, E. M., Hoffecker, J. F., & Gamble, C. (2020). The chronology and function of a new circular mammoth-bone structure at Kostenki 11. *Antiquity, 94*(374), 323–341. https://doi.org/10.15184/aqy.2020.7

55. Boethius, A., Kjällquist, M., Kielman-Schmitt, M., Ahlström, T., & Larsson, L. (2021). Early Holocene Scandinavian foragers on a journey to affluence: Mesolithic fish exploitation, seasonal abundance and storage investigated through strontium isotope ratios by laser ablation (LA-MC-ICP-MS). *PLOS One, 16*(1), e0245222. https://doi.org/10.1371/journal.pone.0245222

56. Binford, L. R. (1980). Willow smoke and dogs' tails: Hunter-gatherer settlement systems and archaeological site formation. *American Antiquity, 45*(1), 4–20. https://doi.org/10.2307/279653

57. Marean, C. W. (2016). The transition to foraging for dense and predictable resources and its impact on the evolution of modern humans. *Philosophical Transactions of the Royal Society B: Biological Sciences, 371*(1698), 20150239. https://doi.org/10.1098/rstb.2015.0239

58. Brown, J. L. (1964). The evolution of diversity in avian territorial systems. *The Wilson Bulletin, 76*(2), 160–169. http://www.jstor.org/stable/4159278

59. Strassmann, J. E., & Queller, D. C. (2014). Privatization and property in biology. *Animal Behaviour, 92*, 305–311. https://doi.org/10.1016/j.anbehav.2014.02.011

60. Walker, M., Johnsen, S., Rasmussen, S. O., Popp, T., Steffensen, J., Gibbard, P., Hoek, W., Lowe, J., Andrews, J., Björck, S., Cwynar, L. C., Hughen, K., Kershaw, P., Kromer, B., Litt, T., Lowe, D. J., Nakagawa, T., Newnham, R., & Schwander, J. (2009). Formal definition and dating of the GSSP (Global Stratotype Section and Point) for the base of the Holocene using the Greenland NGRIP ice core, and selected auxiliary records. *Journal of Quaternary Science, 24*(1), 3–17. https://doi.org/10.1002/jqs.1227

61. Taylor, K. C., Mayewski, P. A., Alley, R. B., Brook, E. J., Gow, A. J., Grootes, P. M., Meese, D. A., Saltzman, E. S., Severinghaus, J. P., Twickler, M. S., White, J. W. C., Whitlow, S., & Zielinski, G. A. (1997). The Holocene-Younger Dryas transition recorded at Summit, Greenland. *Science, 278*(5339), 825–827. https://doi.org/10.1126/science.278.5339.825

62. Yu, Y., He, F., Vavrus, S. J., Johnson, A., Wu, H., Zhang, W., Yin, Q., Ge, J., Deng, C., Petraglia, M. D., & Guo, Z. (2023). Climatic factors and human population changes in Eurasia between the Last Glacial Maximum and the early Holocene. *Global and Planetary Change, 221*, 104054. https://doi.org/10.1016/j.gloplacha.2023.104054

63. Cavalli-Sforza, L. L., & Feldman, M. W. (2003). The application of molecular genetic approaches to the study of human evolution. *Nature Genetics, 33*(S3), 266–275. https://doi.org/10.1038/ng1113

64. Bellwood, P. (2022). *First farmers: The origins of agricultural societies* (2nd ed.). Wiley-Blackwell. https://www.wiley.com/en-au/First+Farmers%3A+The+Origins+of+Agricultural+Societies%2C+2nd+Edition-p-9781119706373

65. Bar-Yosef, O. (1986). The walls of Jericho: An alternative interpretation. *Current Anthropology, 27*(2), 157–162. https://doi.org/10.1086/203413

CHAPTER 8

Spirit in the Sky: Ritual

I.

When Jason motored up to us on the glassy morning water of Abell Point Marina, he could have been Jesus walking on water. He tied his inflatable dinghy to the port side and stepped up. My heart soared when I saw his navy and white sailing shoe contact the deck. *At last, someone knowledgeable on board.*

Jason confidently steered the boat to the fuel dock, where he topped up the fuel and the water. After returning my credit card to me, he maneuvered the boat into its slip in a reversing feat that would cause a panic attack in any normal human. Once the boat was secured to the dock and the engine fell quiet, I was suddenly at a loss. We hadn't planned to be back on land so soon.

Jason asked if we had any face masks with us. In just the few days we had been gone, things on the mainland had apparently changed. New measures to contain coronavirus had been put in place across Queensland. Wearing masks while in public was now mandatory. My husband eagerly answered that yes, we always carry N95 respirators in our backpacks while traveling. I saw him stand taller and lift his shoulders as he spoke, noticing how satisfied he was with his preparedness.

Jason left us and returned to the marina office. We sat for a while in silence. The boat was eerily still, being tied to the dock. It felt more like a stationary house than an ocean-faring vessel. We decided that perhaps we should head into town. What else was there to do? It was mid-morning, and a hot espresso coffee would be amazing.

Airlie Beach is normally a touristy mess of a town, but that morning the streets were empty. We wandered for blocks in what appeared to be a zombie apocalypse. Occasionally we passed a couple of people wearing masks, and we locked eyes with them. *What is going on?* our eyes asked. *Who knows?* theirs answered.

We finally came across a restaurant with five people standing out the front. Through the restaurant's glass exterior, I saw the lights were off and all the chairs were upside down on the tables. Each of the beer taps was covered with a paper bag. No one was inside apart from the barista, who was serving coffee through the window.

My husband ordered his coffee with almond milk and mine with soy. He asked the barista, "Is there somewhere we can sit to drink our coffee?"

The barista paused what she was doing, a bit startled by the question. "No," she answered. "All coffees have to be consumed off site."

When she saw our confusion, her eyes softened, and she took pity on us as if we'd been living under a rock. She explained, "That's the new rules that have come in."

My husband and I looked at each other in indignation. Drinking coffee standing up—what's the point of that?

We grabbed our takeaway coffees and walked with them for a couple of blocks, searching for somewhere along the foreshore to sit. We found a picnic table with a view of the marina. My husband brushed the fallen leaves from the bench, and we took a seat, resting our backs against the table.

I began to contemplate how much I despise the plastic lids that they put on takeaway coffee cups. You'd be better off drinking the beverage through literally anything other than this cheap, flimsy piece of plastic. Sipping coffee through a little hole is nothing at all like the experience of sitting at a marble-topped café table, your legs crossed, waving your foot carelessly as you sip coffee from a glossy, white ceramic cup. The real coffee experience makes you feel like you're in Paris. The cheap takeaway version makes you feel like you're standing in line at Disney World, surrounded by sweaty kids, trying not to barf from the smell of ketchup rising from the hot asphalt.

See, there's a whole ritual to coffee. You have to be sitting in a nice café to enjoy it. The café lends you their table—they welcome you. You take up space in the café as you drink your coffee. The place might be crowded, but your rear end is occupying real estate. Other people may want your table, but they can't have it. It's your right to sit there and drink your coffee. And that makes you feel kind of powerful, like you deserve a place in this world, right here in the middle of society. On the other hand, when you're hurriedly sipping your coffee through a stupid plastic lid, relegated to some park bench on the margins of society, it doesn't have the same magic. It's sacrilege.

Rituals have power because they are familiar. We believe in the familiar. The familiar provides us with a buffer against the cold, uncertain world outside.

Growing up, every year around Thanksgiving, my mom and I would put up the Christmas tree. It was an all-day affair. The first step was to take the artificial tree out of its box, where it had been stored in the basement for the last year. Next came the lights, which were always a tangled mess, and the string of metallic beads. The most delicate part of the ritual began with the hanging of the ornaments. Each ornament had a special meaning. Some were handmade products from our preschool and kindergarten days. Some had returned from family vacations, like the little glass bottle of pink sand from Bermuda. Some ornaments were tokens of our German heritage—a wooden hunter with his green felt hat and feather. Others were gifts from Grandma, including an impossibly heavy metal Statue of Liberty. As we pulled each ornament out of its storage box and brushed off the crinkled packing paper that clung to it, we would smile in recognition, holding the ornament tenderly before finding the perfect spot on the tree. My favorite ornament of all was one that read "Hope," which I would always display prominently near the top.

Now the tree needed a skirt. The felt skirt, which was soft and worn with years of love, would be a welcoming blanket to me and the dog throughout the Christmas season. Next came the placement of the angel. My mom and I are both short, so either my dad or my brother would have to place the angel on top. The dark living room windows created the perfect reflection for our dramatic reveal. With everything in its place, the tree was ready for the final step of the ritual—turning on the lights. We ensured that every other light in the house was off. As the smallest in the family, I crawled behind the tree, plugging in both the Christmas lights and the angel. As the lights burst forth, a hush would overcome us. In an instant, we were reminded that the most important thing is and has always been family.

Ritual does that. It brings us together to perform the familiar and comforts us and strengthens us in the process. Rooted in the traditions of all who have come before us, it gives us the reassurance—the hope—that everything will be okay.

The coffee in my takeaway cup had gone cold. I stared down at the white lid, now stained with coffee. Because of this disposable plastic lid, I would have to wander the streets in search of a recycling bin. Without the sit-down café experience I had expected, I felt uprooted. My coffee-drinking ritual had been thwarted.

On the grass in front of us, I saw three seagulls poking around. Their soft, gray-tipped wings looked as if they had been watercolored. They were not embroiled in a clamorous fight with each other, like I was used to seeing. Instead, their pointy, orange beaks were closed. One headed off to explore the rocks, while another pecked repeatedly at the grass. The third one began tossing his head backwards, as if he was hoping some food would fall from the sky.

The seagulls were probably accustomed to stealing people's fish and chips at the foreshore. It's what they do, day in, day out. But other than the two of us with our coffees, there was no other human in sight. Where were they supposed to get their chips? It would seem that the coronavirus had messed with their ritual as much as it had with ours.

II.

From the outside, the ornate white railing at the roof's edge makes Café Versailles look like a vanilla cake that has been decked for a *quinceañera*. Frilly maroon shades adorn the facade, adding to the institution's old-world appeal. The shade is not purely ornamental, though. It serves as a welcoming respite from the blazing Miami sun. Old Cuban men in their short-sleeved white linen shirts are invited to stand outside the coffee service window and sip their thimble-sized *cafecitos* for as long as they'd like. Here in the shade, you can feel the sea breeze off Biscayne Bay channeled down Calle Ocho, providing a refreshing break from the city's oppressive humidity. Apparently, the gossip is also channeled down the famous street. You can overhear the old men discussing the latest happenings at length in an ongoing lullaby-like hum, punctuated sporadically by the staccato of their crackly, cigar-smoked laughter. Beneath their leather shoes is ground zero of Little Havana, and not one newsworthy event happens without Café Versailles's knowledge.

Drinking coffee at Café Versailles is a ritual, the complexity of which makes it the best coffee in the world. When you order a *café con leche* at Versailles, your coffee comes disassembled. I usually despise "deconstructed" dishes, but at Café Versailles the deconstruction is not about being cool. It's about prolonging the coffee-drinking ritual.

Bread comes first. Served in a paper-lined basket, like a 1950s style diner, the bread is like a baguette but wider, and it has been sliced lengthwise and pressed in a heavy-duty panini press until it is pancake thin. It is slathered with butter to the point where it drips off the end when raised. You begin to break it lengthwise, so now you have four long bread sticks of ideal shape for use in later components of the ritual.

Next comes the ceramic cup, always with an impossibly small handle and always on its own plate, with a little silver spoon resting by its side. The spoon is not for accessing sugar. You don't scoop sugar out of jars at Versailles. You pour it from a glass canister on the table with a free-flow spout. This way you never know how many spoonfuls of sugar you've put in. One? Two? Five? *No importa*—it doesn't matter. We're all going to die one day, and we may as well sit and enjoy our coffee while we can.

But we're getting ahead of ourselves. We don't even have coffee at this stage. All we have is a cup full of hot, whole-fat milk, like you would serve a kitten. You'd be remiss to add sugar to this milk, because you still don't know how sweet the coffee will be.

While you await the arrival of the coffee, you make small talk, like *How has your family been? Your sister is pregnant? They're moving to Kendall?*—nothing too deep, because after all, this is just a preamble to the main course. When the coffee does arrive, and the waiter proudly sets down a steaming silver vessel on your table, you could be forgiven for thinking that a genie is going to emerge, what with the magnitude of hopes that have been pinned on its arrival.

The arrival of the communal espresso vessel marks an important departure point in the ritual. With the coffee here, the really deep conversation can begin. The *I'm not sure whether I should take out a second mortgage* kind of conversation. The *I worry my husband might be cheating on me* kind of conversation. The really juicy stuff, the stuff that we need to pour out of our hearts, just like we need to pour this much awaited espresso.

Pouring the espresso into your cup of milk, you experience a thrill equivalent to what the first alchemists must have experienced. You are performing magic. The dark elixir runs off the end of the spout and hits the fatty, white liquid beneath it, creating all kinds of chaotic patterns as the two mix and the espresso disperses. Finally you are allowed to lift the little silver spoon that has been waiting patiently beside your cup and use it to stir the two. You add espresso until your *café* con leche has reached your desired shade of tan. As you pour, the aroma of the espresso rises into your sinuses and you grin a little, recognizing that other-worldly smell that you're only going to get here, in Miami.

You take the first sip. You may be pleasantly surprised that the Cuban lady at the coffee counter has already sweetened the espresso for you. She's looking out for you, taking care of you. Like a swaddled baby, you relax into a trance. You become lost in conversation, and moments start to break off and disappear like the globules of butter that form on the surface of your *café* con leche as you dip the first buttery toast stick that you prepared earlier into your coffee. You carefully raise the soppy coffee-soaked buttery bread to your mouth. It feels like communion ... but better. Because it is.

You can spend hours drinking coffee and talking at Versailles. No one will raise an eyebrow or dare imply that you're taking too long. Doing so would go against the cultural credo of Cubans, who are never in a rush. Having lived most of their lives in a communist country, waiting in lines for rations, waiting for the person at the bakery to come back from lunch, waiting for someone to get a hold of the right part to fix their '55 Chevy that has broken down yet again, Cubans have no problem waiting.

It always amazes me how after spending just a few days in Miami, carrying out this slow coffee-drinking ritual every day, my own relationship with time is transformed. I think that ritual carries this power. If you give significance to the ritual, it can put you in a different state of mind.

Miami is full of rituals. Sometimes in Little Havana or Little Haiti you'll see people walking down the sidewalk who are dressed in white from head to toe. These individuals are *santeros* or *oungan*, priests of the Cuban Santería and Haitian Vodou religions, respectively. They are kept busy, because in Miami people will go to such priests for blessings or curses to be placed on other people. If a concerned grandmother is praying for her 18-year-old grandson to get into Harvard, she may go to the *santero* to ask that a special rite be done for him, aiding his chances of success. Such rites often involve casting coconut shells, burning incense, or sacrificing animals. By the same token, if you really want to take down your ex-wife or ex-husband, you may go to a *santero* or Vodou priest and ask that a curse be placed on them.

While not everyone in Miami believes in Santería and Vodou, they are also not likely to put the power of their curses to the test. Instead, most agree it's probably best to give the rituals of these Afro-Caribbean religions a wide berth ... sometimes literally.

Tropical Park is a suburban hub for children's birthday parties, for tennis players practicing their backhand, and for joggers of both the casual and devoted kind. The 1 square kilometer park is especially popular among runners because it contains one of the only steep hills in Miami. You wouldn't want to don your brand-new, white running shoes, however, because somewhere on your way up the hill you might trip over a severed cow's tongue.

"Anybody could have tripped on that tongue," said Caroline Galvin, who first encountered it lying in the path at Tropical Park Hill in April 2019. The cow tongue, sliced in half, had been left there as an animal offering by worshippers of Santería, likely as a plea to a god to silence an enemy or gossip. Because the gods of Afro-Cuban Santería are living spirits who must eat, practitioners of the religion offer animal carcasses as a way of feeding them and asking for their assistance.[1]

When the Spanish came to the Americas in the 16th century, they decimated the native populations of Cuba and Haiti.[2] In need of a new labor force to fuel their sugar cane-based imperialist economy, the Spanish brought over slaves from West Africa. These slaves, in turn, brought with them their gods and their rituals. The Yoruba people of western Africa had many rituals to commune with their gods, known as *orishas*. Forced by the Spanish (and later in Haiti, the French) to practice Roman Catholicism, the slaves assigned a West African god, who represented a force of nature, to each of the Catholic saints. This way, to their European masters, it would appear as if the slaves were worshipping the saints, when in reality they were still performing rites and offering sacrifices to their *orishas*.[3]

The tongue is not the first animal sacrifice to appear on Tropical Park Hill. In the past, joggers have confronted parts of dead chickens, roosters, pigeons, goats, and pigs, right there on the running path. Galvin reported, "I found a bag with a rooster's feet sticking out of it and plastic containers full of blood. My neighbors found a tongue nailed to a tree."[1]

Such ritualistic symbols are scattered all over the metropolis. If you go to the courthouse in downtown Miami, you may stumble upon a pig's head deposited on the steps as a plea for justice. If you go for a sail on Biscayne Bay, you may have to dodge a bobbing watermelon, a sacrifice to Yemaya, goddess of the sea and motherhood.[1]

Although they may be repulsive to your average jogger or weekend picnicker, no one can stop the animal sacrifices. In the 1993 U.S. Supreme Court case of *Church of the Lukumi Babalu Aye, Inc. v. City of Hialeah*, the Court ruled in favor of the Santería church, with Justice Anthony Kennedy concluding that Hialeah's city ordinances preventing religious animal sacrifices violated the Free Exercise Clause of the First Amendment because they were designed to persecute or oppress a religion or its practices.[4] Beyond constitutional freedoms, these rituals also give a sense of power to marginalized immigrant communities, ensuring that they still have a voice in a country where they often do not speak the primary language or fit within the dominant culture.

Looking at how the Yoruba people managed to hold onto their religion even in the face of Spanish oppression, and how Café Versailles has held onto its old-world approach to drinking coffee even in the midst of fast-and-furious Miami, we can see how vital ritual can be to a human's sense of agency.

Mercedes Sandoval, an anthropologist at Miami Dade College said about the cow tongue incident at Tropical Park, "We see these remains and we ask, what is this type of religion doing in 21st century Miami?" Contemplating the ritual's deeper significance, Sandoval said, "Magic can give you the feeling that you are gaining control."[1]

As long as the *orishas* get fed, there is a sense that things are under control. But what happens, dare I ask, if the gods don't get fed?

III.

Watching an animal who is expecting to be fed but is not fed has got to be one of the most curious experiments in animal behavior. Anticipation of food is an experience that has been used in many studies of animals. It is easy to replicate and control: Over several days, get a female sheep used to being fed at the same time and place every day, and then one day, feed her an hour later. Watch what happens. More often than not, she becomes very distressed. Why? Because you're screwing with her expectations.

Expectations are potent for animals. If we are expecting food (or anything we really like), we get antsy about its arrival. Our expectations have dramatic effects on how we feel.[5] Although there are many definitions of animal welfare in the scientific literature, as far as I know, the French government's definition of animal welfare is the only one to take into account an animal's expectations.[6] Our environment certainly influences how we feel, the French agree, but so do our own expectations.

If you've ever observed children as they wait in line to get ice cream, you understand the power of expectations. Children in this situation can't possibly hold still. The time between knowing they're going to get ice cream and actually putting the ice cream in their mouths is an eternity for kids. In animal behavior terms, this time of waiting is known as the "anticipatory period."[7] What an animal chooses to do during the anticipatory period will in no way make the food arrive any faster. And yet many animals, including human children, will perform funny-looking behaviors during the anticipatory period. I've seen kids jump up and down when they were standing in line for ice cream. The jumping doesn't make the ice cream show up any faster, and yet they jump. This is a common anticipatory behavior—anything that makes you feel like you're getting closer to the food. The point is, the behavior is useless; it doesn't get you to the food any faster.

Nikolaas Tinbergen was a Dutch biologist who was fascinated by the spontaneous behaviors that animals perform, behaviors that appear to arise from nowhere. Why were they performing these behaviors? After

WWII, Tinbergen moved to Oxford, England, and to answer his questions, he turned to watching the closest subjects he could find—seagulls.[8]

The European herring gull (*Larus argentatus*) inhabits most of the frigid shores of northern and western Europe. Some herring gulls migrate south for the winter, but many live year-round on the shores of the North Sea, including the coasts of England.

From 1950, Tinbergen and his students recorded a number of behavioral movements that the gulls practiced, some of which almost resembled "rituals." For instance, in the "long call" behavior, a European herring gull would stretch his head forward and upwards and let out one hoarse call. Then he would jerk his head downwards, letting out one or two muffled, high-pitch calls. Following this, he would throw his head back with a jerk, beginning a series of loud calls. With each loud call, he would lower his head a bit, like a lawn sprinkler, until he ended back at horizontal. Each time a gull did this sequence, they did it in nearly the same way.[9]

Tinbergen and his team noted many other ritualistic displays in gulls, including "the swoop-and-soar," "jabbing," "choking," "head tossing," and "pecking into the ground." For many of these behaviors, there was no obvious function. They seemed to serve no survival purpose.[9]

American psychologist B.F. Skinner had seen similarly purposeless behavior in pigeons just a couple of years earlier. Skinner noticed that the pigeons at Indiana University performed funny-looking behaviors while they were anticipating food. In one experiment, he placed the pigeons in cages and drastically reduced their food intake, causing them to lose 25% of their normal body weight. Then he began to present food to the pigeons for brief periods—just 5 seconds. The presentation of the food was set to a clock—appearing every 15 seconds, regardless of what the pigeons were doing at the time.[10] The pigeons were hungry, and here was this food appearing, but only for a few seconds.

Skinner noticed that the pigeons began to offer random behaviors to try to make the food come faster. Two of the birds rocked their heads and bodies from side to side in a pendulum motion. One of the birds turned counterclockwise about the cage, making two or three turns between food appearances. Another bird developed a head tossing behavior, whereby she placed her head down and then lifted it suddenly. And another bird made pecking movements towards the floor. These movements were repeated in the same part of the cage, generally in the same orientation, multiple times in rapid succession.[10]

As long as the intervals between food appearances were short—10, 15, or 20 seconds—each pigeon would continue to offer his or her ritualistic behavior. Even when the intervals were longer, it took 10 to 15 minutes of not receiving food before the birds finally gave up and stopped offering the behaviors. Skinner compared the pigeons' ritualistic behavior to the "luck-enhancing" behaviors of a bowler who has released the ball down the alley but continues to twist and turn his arm, as if he were still controlling the ball. Skinner described the pigeons' strange movements as "a sort of superstition."[10]

Skinner suggested that by presenting food regularly, regardless of what behaviors the pigeons were offering, he had caused the birds to develop a false association. They had made an accidental connection between their rituals and a favorable consequence.[10] They thought that the food was appearing because of the rituals they were carrying out, when in reality the food's appearance was completely out of their control.

Skinner's brand of American determinism had always been regarded with suspicion by European scientists, who believed that animals could not be reduced to simple behavior-producing machines. From the early 1950s, Tinbergen and his colleague Konrad Lorenz began to hypothesize that the birds' seemingly non-functional behavioral displays might be serving a function after all. The birds might be communicating to each other with the behaviors, using them as social signals.[11]

Tinbergen's observations of seagulls provided support for this social signaling hypothesis. He found that some of the gull's ritualistic behaviors, including jabbing, choking, and pecking into the ground, were almost always performed in situations where there were intruders encroaching on a gull's territory. The effect of these threat display behaviors was to make the intruding birds move away or stop their advance. Meanwhile, another set of behaviors, including head tossing, facing away, and the hunched and upward choking postures, achieved a different aim—they brought other birds closer, either for the purpose of mating or for feeding.[9] When a gull chick was asking for food from a parent, head tossing stimulated the parent to regurgitate the food. During mating, head tossing served the purpose of drawing a bird of the opposite sex closer.[9] Tinbergen suggested that when gulls perform the seemingly random behavior of head tossing in different contexts, the behavior is not just an accident, as Skinner had suggested. Instead, he believed that ritualistic behaviors are related to animals' innate motivations to access critical resources, like food or mates.[9]

Tinbergen's findings were supported when ethologists William Timberlake and Gary Lucas repeated Skinner's experiment at Indiana University in 1985. Unlike Skinner, they found remarkable consistency in the patterns of behavior offered by the pigeons. Almost all of the birds carried out some sort of behavior directed at the wall where the food usually appeared. This included head bobbing, pendulum motions of the neck, bumping the panel with the breast, and scratching the wall with their feet. Timberlake and Lucas suggested that the pigeons' behaviors were not random or superstitious at all but were species-typical patterns of behavior that were commonly associated with the anticipation of food in the wild.[12]

Today our understanding of the functional significance of ritualistic behaviors in birds has been advanced even further by the millions of birds from hundreds of species held in captive settings. In zoos, aviaries, chicken coops, and household cages, it is not uncommon for birds to offer highly repetitive behaviors that almost seem like compulsive rituals for the birds. Animal behavior specialists refer to them as "stereotypies" or "abnormal repetitive behaviors."[13]

Abnormal repetitive behaviors have been documented extensively across mammals in captive settings and to a lesser extent in captive birds. In captive birds, commonly reported abnormal repetitive behaviors include feather pecking, pacing along defined routes, somersaulting, pecking at a certain spot or object, and spinning, rocking, or twirling motions of the body or head.[13] The more predictable the environment and the appearance of food, the more the perceived control, and the more frustrating it will be if the animal subsequently loses that control.[14]

Although the function of abnormal repetitive behaviors is not readily apparent, it is generally supposed that they calm the animal or help the animal cope with stress.[13, 15] This is supported by the observation that providing animals with drugs that affect the dopaminergic and serotonergic pathways results in a reduction in these behaviors.[16] We know that for humans, repetitive behavior can be calming, such as the repeating of mantras by Buddhist monks.[17] It could be that abnormal repetitive behaviors help an animal regain a sense of control, thereby reducing the animal's anxiety.

The seagulls at Airlie Beach were accustomed to being fed chips every day when people ate fish and chips on the foreshore. Now, because of coronavirus, they could no longer access their preferred resource. So, they did what any reasonable animal would do—they tried to take control of the situation by offering whatever behaviors they could, as fruitless as they were.

IV.

To some people, religious ceremonies look as silly as a pigeon who is turning around in circles, but in the chaotic context of our evolution, human rituals aimed at the supernatural are not that unreasonable.

If there's absolutely no value in performing rituals aimed at the supernatural, then why do we see them in nearly every human culture in the world? Pleasing the gods or spirits is no easy task—it involves huge investments of time and resources to carry out rituals, build temples, and offer sacrifices. By definition, rituals involve predetermined sequences of behaviors, which must be performed in a rigid, formal, or repetitive way.[18] For humans, these sequences draw heavily on our limited energy. In order to persist, then, surely the practice of rituals must confer some sort of adaptive advantage for human survival.

The advantage is certainly not an obvious one. By its very nature, a ritual produces no practical result that can be readily observed in the external environment. This is what distinguishes a ritual from a habit. A habit can have an instrumental purpose of getting something done, but a ritual lacks instrumental purpose. Despite its impracticality, ritualistic behavior does have great value to the people who practice it. This is because rituals are always embedded in some sort of larger cultural system of symbolism and meaning, and these systems of meaning, created by humans, are what give rituals their value.[18]

Inferring systems of meaning from the prehistorical record is understandably difficult—beliefs are rarely recorded in stone. That being said, some of the earliest indications of rituals practiced by humans come from materials that humans collected and used purely for their symbolic value.

Take, for instance, the 57 pieces of pigment that archaeologists have discovered at Pinnacle Point, a cave overlooking the Indian Ocean at the tip of South Africa.[19] The pigment pieces were found together with *Homo sapiens* remains and tools, which are around 164,000 years old.[19] Chemists have classified the pigment as "red ocher," a fine-grained Earth material that is rich in iron oxides and reddish in color. Most of the pigment pieces at Pinnacle Point are ground on one surface, showing their modification and use by humans. Interestingly, the very-red pigments are the ones that have been used the most. This suggests that the humans who used them might have had a preference for the reddest of red pigments.[19]

The preferential use of the reddest ochers is not unique to Pinnacle Point. It has been found at other South African sites, such as Blombos Cave, Rose Cottage, and Sibudu, where early humans lived and used pigments. From about 160,000 years ago, red ocher appears in almost every South African cave and rock shelter where humans lived.[20] What is interesting for our understanding of rituals, though, is that red ocher has no obvious function for human survival.[21] It's apparently a useless material. So what were early humans doing with it?

There is good reason to believe that for our ancestors, the color red held symbolic value. Red possesses special attention-grabbing properties for primates, likely because of our of association of red with ripe fruits, blood, and the color of flushed skin during an emotional response.[20, 21] Although other natural pigments, like yellow and black, were readily available in their environment, it appears that early humans preferentially collected, processed, and used red ocher.[21] Between 160,000 and 40,000 years ago, the use of red ocher became habitual in *Homo sapiens* societies.[21] It has been found in nearly every place where humans roamed. Many sites with *Homo sapiens* remains have over 1000 pieces of red ocher pigment.[21] Some of these pieces have been engraved with geometric shapes for what could have been symbolic reasons,[20] while other quantities of ocher have been pulverized using specialized tools and mixed with heated animal fat into a thick, red paint. At Blombos Cave in South Africa, for instance, humans built an entire ocher processing workshop and stored the red paint in dedicated shell containers.[22] Early humans likely applied red ocher to their skin, as well as tools and other surfaces.[21]

Some of the oldest examples of jewelry (or "personal ornaments", as archaeologists call them) were painted red. Inside Qafzeh cave in Israel, scientists have found a collection of 92,000-year-old seashells, some of which still have traces of red ocher. The seashells appear to have been collected from the Mediterranean Sea, 35 km away. Only shells with holes in them appear in the cave, and in some of the shells, the holes have been intentionally widened through percussion. There is evidence of friction on one side of each hole, indicating the shells have been hung from a string. The team who studied the shells concluded that they were likely strung as beads and used as personal ornaments, and they were either deliberately painted red or had rubbed against the skin of someone painted with ocher.[23] Other seashells with holes, also bearing traces of red ocher and wear patterns from being suspended, have been found at Grotte des Pigeons in Morocco from 82,000 years ago.[24] The use of shell beads as personal ornaments is generally considered to be robust evidence of symbolic culture among humans.[20]

Taken together as a body of evidence, these materials suggest that as early as 160,000 years ago, humans were using red ocher with purpose—selecting for the most intense red, modifying the pigment through specialized processing, sacrificing valuable resources like animal fat to make the ocher into paint, and engraving pieces of ocher with geometric emblems. What our ancestors once believed about painting themselves red or wearing a string of painted shells is a matter of speculation. We may never understand their systems of meaning in any detail, but it does suggest that long ago, humans in Africa were regularly engaging in energy-intensive behaviors that went beyond just practicality. In other words, some sort of symbolic ritual was happening.

Anthropologist Ian Watts has argued that if prehistoric humans were participating in group rituals some 100,000 years ago, then we would expect these rituals to leave a "loud archaeological signature."[25] In other words, we should find a lot of physical traces left by early humans who were repeatedly visiting the same locations, year after year, engaging in the same complex patterns of behavior, despite the difficulty and lack of instrumental purpose. It turns out that, as Watts predicted, we can spot this loud archaeological signature from very far away.

Jutting up from the flat landscape of the Kalahari Desert in northwestern Botswana is a handful of rocky peaks, the highest of which is 400 m tall. These hills are the only hills for 100 km in any direction, and they have stood as a beacon for *Homo sapiens* for the last 100,000 years. The Tsodilo Hills contain over 4,500 rock paintings from those years, making it one of the oldest and highest concentrations of rock art in the world.[26] These hills also act as a time capsule of human beings' ancient communion with the spirit world.

The artwork inside one narrow, difficult-to-access cave in Tsodilo Hills is particularly revealing. Rhino Cave sits high on the northernmost face of the second tallest hill. To get there, you have to ascend to the ridge, climb over several large boulders, and then squeeze yourself through a raised, narrow crawl space before dropping a considerable distance to the cave's floor. Inside there is hardly any sunlight. Not a speck of vegetation grows on the powdery floor. The cave is regularly flooded during heavy rains, which makes it unsuitable for long-term occupation by humans.[26] But for a short visit, say to contact the spiritual realm, Rhino Cave is the perfect spot.

The interior of Rhino Cave is long and narrow, varying from 1.25 m to 5 m wide, but the cave doesn't feel claustrophobic because of its high ceiling.[26] It would hold around 40 people if they were all standing and didn't need to move around much, but allowing for movement, you'd expect no more than 15 people to fit. There is evidence that humans have been visiting this cave for at least the last 70,000 years, although there's no indication of anyone having lived there.[26] Both of the cave's long, main walls are decorated. The northern wall has a series of rock paintings, including a meter-long rhinoceros painted in white, a giraffe painted in red, and several geometric emblems over and around the animals, the most repeated one being a circle enclosing a grid.[27] The opposite wall, which is 7 meters long and 2 m high, has had over 300 grooves and cups carved out of it so that it resembles a long lattice.[26] Or, viewed in different light, a larger-than-life animal.

The carved wall inside Rhino Cave is dark for most of the year. The exception is during the middle of winter, when for a few hours in the late afternoon, a narrow arc of sunlight enters through a small hole in the ceiling. The arc of sunlight hits the wall directly and illuminates the carvings.[26] At this time, the hundreds of depressions in the wall become the scales on the body of a glowing, 7 meter snake. A similar effect is produced by the light of a flame. Archaeologist Sheila Coulson carried out a series of nighttime experiments in which she used candles to observe the effect of flickering light on the wall. Firelight emphasizes a large indentation and a crack in the rock, forming the snake's eye and mouth. What's more, the flickering light across the carvings creates the illusion that the snake's body is moving.[26] Directly in front of the carved wall, Coulson and her colleagues found 88 stone points, most of which had been manufactured from materials sourced from far away and intentionally burned, rendering them brittle and useless but more brilliant in color. They concluded that the only purpose of the burnt points could have been for rituals and that these points, together with the undulating snake wall and the rock paintings, point to use of the cave as a site for rituals over many years. That is, Rhino Cave provides a loud archaeological signature of collective rituals by *Homo sapiens* during the Middle Stone Age, some 70,000 years ago.[26]

Caves, with their isolation, darkness, and acoustic properties, are the perfect places in which to carry out mind-altering rituals. Watching a fire is mesmerizing no matter where you are, but inside a cave, the flickering of a fire is amplified against the walls, stalactites and stalagmites, making the place look otherworldly. Even the fire itself can induce altered states of consciousness in visitors to the cave. Archaeologist Yafit Kedar and colleagues simulated the effect of one burning wood torch inside caves of varying dimensions and studied the dispersal of smoke and oxygen concentrations over time. They found that within the typical narrow cave painted by early humans, the oxygen level dropped from its natural 21% to below 18% after just 15 minutes of wood burning. This decrease would be enough to cause mild hypoxia in humans. In caves with low ceilings, the oxygen concentration fell to 11% after 2 hours, which would induce severe hypoxia. Hypoxia is what happens when the brain is starved of oxygen, and it is associated with drowsiness, euphoria, and hallucinations. Hypoxia also increases dopamine levels in the body, which can cause dreams, hallucinations, and out-of-body experiences, like the sensation of flying. The darkness of the cave environment would have exacerbated these altered states of consciousness, producing visions of spiritual beings emerging from the recesses of the cave.[28]

Some of the earliest human cave art suggests that our ancestors came to caves in search of animals — not so much physical animals like the game species that they regularly hunted, but powerful spirit animals, like snakes, rhinos, and lions, who could act as guides to the underworld.[29] Cross-cultural research reveals that in almost all hunter-gatherer societies around the world where traditional religions are practiced, shamans use rituals to induce altered states of consciousness in an effort to commune with spirit animals.[30] These rituals often involve some form of fasting, dehydration, fatigue or sleep deprivation, self-inflicted pain, blood loss, or sensory deprivation to reach ecstatic states. The human mind can be further altered through the use of drugs, hyperventilation, rhythmic light flashes, rhythmic music and dance, or repetitive chants.[31] Early humans, like the visitors to Rhino Cave, depicted their visions of spirit animals on the walls of caves where they practiced their rituals. They might have believed that the spirit animals drawn out through shamanic rituals would help them to find game, bring rain, increase fertility, heal people, or maintain their balance with nature.[30]

Apart from inducing hallucinations, caves are also excellent places for spiritual rituals because of their exclusivity. Caves are tight places. Given that much of the archaeological evidence of rituals by early *Homo sapiens* is found deep within caves, where caves are at their narrowest, it is likely that only a select few people participated in such rituals. Caves also act like a big megaphone, so any chants spoken or songs bellowed within a cave would have echoed and been broadcast to all those outside the cave. People might have assumed that supernatural things were going on inside caves and that those people inside the cave had come in contact with the divine.[31] In this way, ritual might have served a secondary function of social signaling for our ancestors, as Tinbergen suggested ritualistic behavior did for seagulls. The performance of costly rituals by shamans and a select few people who entered Rhino Cave, for instance, might have signaled to others that these individuals had a privileged connection with the spiritual realm, or at the very least, that they were more devoted to the wellbeing of the community. Anthropologist Brian Hayden argues that, in this way, cave rituals could have paved the way for the emergence of an elite class.[31]

According to many anthropologists, the evolution of religion is just "a supernatural extension of human social life."[32] When humans take part in a ritual, it reaffirms our position in the group and enhances our connections with fellow group members by advertising signals of loyalty and trust.[18] Evolutionary psychologist Matt Rossano has suggested that the practice of group rituals by humans might have evolved as a kind of "social security." Rituals protect humans against the threat of future uncertainty and reduce our feelings of helplessness. At the same time, rituals can make human groups stronger by enhancing group members' commitment to communal values, enlisting the goodwill of others (both human and supernatural), and building networks of social support.[33]

According to Rossano, "Rituals address two problems perennially confronted by human communities—combating social entropy (disintegration) and quelling future anxieties. Ultimately, these two concerns are one and the same: quelling future anxieties, because group disintegration is a harbinger of future uncertainty."[33]

Performing religious rituals could have been our ancestors' way of seeking reassurance that game would appear, an attempt to control the uncontrollable. Perhaps our ancestors relied on shamans or spiritual leaders to participate in religious rituals, in the hopes that supernatural forces would bring more reliable food. As with Skinner's pigeons, when a surplus of game appeared one year, the effectiveness of the ritual was reinforced. In years when game animals were scarce, our ancestors might have believed they had done something to anger the animal spirits, in which case more ritual sacrifices would be required.

While rituals might have made early human societies more cohesive on the one hand, they might have also exacerbated inequalities among individuals. If only a few individuals participated in mind-altering rituals within the narrow confines of caves, then perhaps these individuals gained greater respect, status, or power within their communities. Building upon the **Illusion of Security** and the feeling that they could somehow influence the forces of nature through ritual, our ancestors might have paved the way for a second illusion—the **Illusion of Superiority**. They might have begun to believe that certain people are actually more valuable than others, due to their connections with the supernatural world. Or in the words of George Orwell, "All animals are equal, but some animals are more equal than others."[34]

As rituals continued and became more embedded in cultures, they likely brought powerful social advantages to our ancestors. The act of participating in rituals might have enabled early humans to prove their commitment and trustworthiness within their groups, as well as establish peaceful alliances with neighboring groups. Successful alliance building at a regional level would become a major advantage later in our history, once our ancestors sought to expand their empires.

CHAPTER 8 REFERENCES

Free Resources

Marean, C. W. (2015, August 1). *How* Homo sapiens *became the ultimate invasive species.* Scientific American. https://www.scientificamerican.com/article/how-homo-sapiens-became-the-ultimate-invasive-species/

Surugue, L. (2018, February 22). *Why this Paleolithic burial site is so strange (and so important).* SAPIENS. https://www.sapiens.org/archaeology/paleolithic-burial-sunghir/

Further Reading

Hayden, B. (2003). *Shamans, sorcerers, and saints: A prehistory of religion.* Smithsonian. https://smithsonianbooks.com/store/anthropology-archaeology/shamans-sorcerers-and-saints/

Rossano, M. (2010). *Supernatural selection: How religion evolved.* Oxford University Press. https://global.oup.com/academic/product/supernatural-selection-9780195385816

References

1. Paterson, L. (2019, April 18). Watch your step in Miami. There might be a severed cow's tongue on your running path. *Miami Herald.* https://www.miamiherald.com/news/local/community/miami-dade/article228861589.html

2. Fernandes, D. M., Sirak, K. A., Ringbauer, H., Sedig, J., Rohland, N., Cheronet, O., Mah, M., Mallick, S., Olalde, I., Culleton, B. J., Adamski, N., Bernardos, R., Bravo, G., Broomandkhoshbacht, N., Callan, K., Candilio, F., Demetz, L., Carlson, K. S. D., Eccles, L., ... Reich, D. (2021). A genetic history of the pre-contact Caribbean. *Nature, 590*(7844), 103–110. https://doi.org/10.1038/s41586-020-03053-2

3. Murrell, N. S. (2010). *Afro-Caribbean religions: An introduction to their historical, cultural, and sacred traditions.* Temple University Press. https://muse-jhu-edu.ap1.proxy.openathens.net/book/45938

4. Church of the Lukumi-Babalu Aye v. Hialeah, 508 U.S. 520. (1993). https://www.uscourts.gov/educational-resources/educational-activities/exercise-religious-practices-rule-law

5. Anderson, C., Yngvesson, J., Boissy, A., Uvnäs-Moberg, K., & Lidfors, L. (2015). Behavioural expression of positive anticipation for food or opportunity to play in lambs. *Behavioural Processes, 113*, 152–158. https://doi.org/10.1016/j.beproc.2015.02.003

6. French Agency for Food, E. and O. H. & S. (2018, April 25). ANSES proposes a definition of animal welfare and sets the foundation for its research and expert appraisal work. *ANSES.* https://www.anses.fr/en/content/anses-proposes-definition-animal-welfare-and-sets-foundation-its-research-and-expert

7. Anderson, C., von Keyserlingk, M., Lidfors, L., & Weary, D. (2020). Anticipatory behaviour in animals: A critical review. *Animal Welfare, 29*(3), 231–238. https://doi.org/10.7120/09627286.29.3.231

8. Beer, C. (2020). Niko Tinbergen and questions of instinct. *Animal Behaviour, 164,* 261–265. https://doi.org/10.1016/j.anbehav.2019.08.005

9. Tinbergen, N. (1960). Comparative studies of the behaviour of gulls (*Laridae*): A progress report 1. *Behaviour, 15*(1–2), 1–69. https://doi.org/10.1163/156853960X00098

10. Skinner, B. F. (1948). "Superstition" in the pigeon. *Journal of Experimental Psychology, 38*(2), 168–172. https://doi.org/10.1037/h0055873

11. Tinbergen, N. (1953). *The herring gull's world: A study of the social behaviour of birds.* Collins. https://books.google.com.au/books/about/The_Herring_Gull_s_World.html?id=9fM5AAAAMAAJ&redir_esc=y

12. Timberlake, W., & Lucas, G. A. (1985). The basis of superstitious behavior: Change contingency, stimulus substitution, or appetitive behavior? *Journal of the Experimental Analysis of Behavior, 44*(3), 279–299. https://doi.org/10.1901/jeab.1985.44-279

13. Mellor, E., Brilot, B., & Collins, S. (2018). Abnormal repetitive behaviours in captive birds: A Tinbergian review. *Applied Animal Behaviour Science, 198,* 109–120. https://doi.org/10.1016/j.applanim.2017.09.011

14. Bassett, L., & Buchanan-Smith, H. M. (2007). Effects of predictability on the welfare of captive animals. *Applied Animal Behaviour Science, 102*(3–4), 223–245. https://doi.org/10.1016/j.applanim.2006.05.029

15. Mason, G., & Latham, N. (2004). Can't stop, won't stop: Is stereotypy a reliable animal welfare indicator? *Animal Welfare, 13*(S1), S57–S69. https://doi.org/10.1017/S096272860001438X

16. Langen, M., Kas, M. J. H., Staal, W. G., van Engeland, H., & Durston, S. (2011). The neurobiology of repetitive behavior: Of mice.... *Neuroscience & Biobehavioral Reviews, 35*(3), 345–355. https://doi.org/10.1016/j.neubiorev.2010.02.004

17. Gao, J., Leung, H. K., Wu, B. W. Y., Skouras, S., & Sik, H. H. (2019). The neurophysiological correlates of religious chanting. *Scientific Reports, 9*(1), 4262. https://doi.org/10.1038/s41598-019-40200-w

18. Hobson, N. M., Schroeder, J., Risen, J. L., Xygalatas, D., & Inzlicht, M. (2018). The psychology of rituals: An integrative review and process-based framework. *Personality and Social Psychology Review, 22*(3), 260–284. https://doi.org/10.1177/1088868317734944

19. Marean, C. W., Bar-Matthews, M., Bernatchez, J., Fisher, E., Goldberg, P., Herries, A. I. R., Jacobs, Z., Jerardino, A., Karkanas, P., Minichillo, T., Nilssen, P. J., Thompson, E., Watts, I., & Williams, H. M. (2007). Early human use of marine resources and pigment in South Africa during the Middle Pleistocene. *Nature, 449*(7164), 905–908. https://doi.org/10.1038/nature06204

20. Henshilwood, C. S., d'Errico, F., & Watts, I. (2009). Engraved ochres from the Middle Stone Age levels at Blombos Cave, South Africa. *Journal of Human Evolution, 57*(1), 27–47. https://doi.org/10.1016/j.jhevol.2009.01.005

21. Dapschauskas, R., Göden, M. B., Sommer, C., & Kandel, A. W. (2022). The emergence of habitual ochre use in Africa and its significance for the development of ritual behavior during the Middle Stone Age. *Journal of World Prehistory, 35*(3–4), 233–319. https://doi.org/10.1007/s10963-022-09170-2

22. Henshilwood, C. S., d'Errico, F., van Niekerk, K. L., Coquinot, Y., Jacobs, Z., Lauritzen, S.-E., Menu, M., & García-Moreno, R. (2011). A 100,000-year-old ochre-processing workshop at Blombos Cave, South Africa. *Science, 334*(6053), 219–222. https://doi.org/10.1126/science.1211535

23. Bar-Yosef Mayer, D. E., Vandermeersch, B., & Bar-Yosef, O. (2009). Shells and ochre in Middle Paleolithic Qafzeh Cave, Israel: Indications for modern behavior. *Journal of Human Evolution, 56*(3), 307–314. https://doi.org/10.1016/j.jhevol.2008.10.005

24. Bouzouggar, A., Barton, N., Vanhaeren, M., d'Errico, F., Collcutt, S., Higham, T., Hodge, E., Parfitt, S., Rhodes, E., Schwenninger, J.-L., Stringer, C., Turner, E., Ward, S., Moutmir, A., & Stambouli, A. (2007). 82,000-year-old shell beads from North Africa and implications for the origins of modern human behavior. *Proceedings of the National Academy of Sciences, 104*(24), 9964–9969. https://doi.org/10.1073/pnas.0703877104

25. Watts, I. (2009). Red ochre, body painting, and language: Interpreting the Blombos ochre. In *The cradle of language* (pp. 62–92). Oxford University Press. https://doi.org/10.1093/oso/9780199545858.003.0004

26. Coulson, S., Staurset, S., & Walker, N. (2011). Ritualized behavior in the Middle Stone Age: Evidence from Rhino Cave, Tsodilo Hills, Botswana. *PaleoAnthropology*, 18–61. https://doi.org/10.4207/PA.2011.ART42

27. Robbins, L. H., Murphy, M. L., Campbell, A., & Brook, G. A. (1996). Excavations at the Tsodilo Hills Rhino Cave. *Botswana Notes & Records, 28*(1), 23–53. https://journals.co.za/doi/pdf/10.10520/AJA052550590_398

28. Kedar, Y., Kedar, G., & Barkai, R. (2021). Hypoxia in Paleolithic decorated caves: The use of artificial light in deep caves reduces oxygen concentration and induces altered states of consciousness. *Time and Mind, 14*(2), 181–216. https://doi.org/10.1080/1751696X.2021.1903177

29. Lewis-Williams, D. J., & Clottes, J. (1998). The mind in the cave — the cave in the mind: Altered consciousness in the Upper Paleolithic. *Anthropology of Consciousness, 9*(1), 13–21. https://doi.org/10.1525/ac.1998.9.1.13

30. Winkelman, M. (2002). Shamanism and cognitive evolution (with comments). *Cambridge Archaeological Journal, 12*(1), 71–101. https://doi.org/10.1017/S0959774302000045

31. Hayden, B. (2003). *Shamans, sorcerers, and saints A prehistory of religion*. Smithsonian. https://smithsonianbooks.com/store/anthropology-archaeology/shamans-sorcerers-and-saints/

32. Rossano, M. J. (2007). Supernaturalizing social life. *Human Nature, 18*(3), 272–294. https://doi.org/10.1007/s12110-007-9002-4

33. Rossano, M. J. (2020). Ritual as resource management. *Philosophical Transactions of the Royal Society B: Biological Sciences, 375*(1805), 20190429. https://doi.org/10.1098/rstb.2019.0429

34. Orwell, G. (1945). *Animal farm*. Secker & Warburg. https://books.google.com.au/books?id=My-TZvEZZFIC&source=gbs_book_other_versions

CHAPTER 9

Eve of Destruction: Imperialism

I.

The liquor store at Airlie Beach was only a few steps away from Woolworths. My husband wheeled our trolley over the brick footpath, and I followed close behind him. While on land, we took the opportunity to stock up on tortillas, beer, and other essentials. Everyone in the shopping center was wearing a face mask. There was a queue outside the liquor store—everyone was stocking up. The people queuing had arranged themselves equidistant from each other, about one meter apart, like king penguins sitting on their eggs.

Normally my husband and I like to take our time, perusing the craft beer section for hitherto unexplored varieties, but that day our beer browsing was cut short. There was a sign taped to the door: only two customers allowed inside.

Four people were in front of us in line, which meant two more bouts of customers before it would be our turn. My husband turned and peered at me over his mask with restless eyes. He wasn't the only one who hated standing in line. I looked around at the others, who were playing on their phones and staring mindlessly into the distance. Inside they must have been seething too, but they were trying to keep it together as functioning members of society.

When it was finally our turn, we made a bee line for the craft beer fridges at the back. Nothing was on sale, really, so we settled on two of our favorites—Balter XPA and Black Hops IPA. My husband set down our two four packs of beer on the counter, and I added a chilled bottle of Viognier that I found at the end. The cashier scanned them and put them into a cardboard wine box. After placing them inside, he immediately reached for a bottle of hand sanitizer next to the cash register and pumped a dollop into his palm. It struck me as rude, as if we were somehow dirty.

The next two people in line watched us through the glass door. They saw my husband pull out his credit card to pay, signaling to them that it was almost their turn. The payment went through, and the cashier said, "Thanks. Have a great day." As he pushed the cardboard box across the counter, I felt like a prisoner in the mess hall, having received my allotted scoop of porridge.

The next morning, we left the marina as early as we could. We had to escape the dismal, state-imposed prison before someone told us we couldn't. As we left the marina and entered the channel, we saw many other boats joining the exodus. Normally fleets of sailboats exiting a marina at once would indicate the start of a race. But that day, there were no sailing crews wearing colorful matching shirts. It was just weathered, old sailors, escaping to sea.

We decided to sail to an anchorage at Stonehaven, on the northwest side of Hook Island. It wasn't far, so we would be there in a matter of hours. We had heard there was good snorkeling there, and we needed a refreshing snorkel after our stale experience in Airlie Beach.

The wind was good, so we put out both the sails. We zoomed along on a beam reach. It felt sneaky being out on the water when the rest of Queensland was under such strict restrictions. It was like we were skipping school.

We hadn't had our sails up for long before we needed to drop them again to approach the anchorage. Although it was the middle of the day, seven other boats were already there. I searched the water's surface for a free mooring ball that we could take. Only one remained, near the shallowest part of the cove. This last remaining mooring was not ideal—I would have preferred deeper water. In the Whitsundays, it seemed there were rock-hard reefs everywhere.

As my husband flaked the mainsail, I steered us in with caution, keeping one eye on the depth meter. He moved to the front of the boat, boat hook in hand. We approached the mooring ball, and I could see our depth dropping, but we still had just over a meter of water under the keel. I slowed the boat right down, kicking it into reverse to stop our forward motion. From the front of the boat, my husband used his extended arms to signal to me, directing me to steer to the right or the left. At this stage only he could see the mooring ball.

I saw him reach down with the boat hook and pull up a bright orange plastic ball. Attached to the ball was a dripping, thick rope, which was covered with small barnacles near the base. He put the rope over the cleat at the bow and then tied another rope on top of it to secure it in place. He dropped the orange ball back onto the water's surface right in front of our bow. I killed the engine.

Looking at the surrounding boats, I noticed that most of them were larger than us. One was a shiny black and white superyacht. It looked big enough to have a swimming pool on board. No one could be seen inside its black reflective windows. I was curious about what kind of anal sailors felt the need to get to the anchorage by noon and claim their stake to Stonehaven's moorings.

Someone once told me about the resorts in Mallorca, off the coast of Spain, where Europeans go on vacation. Supposedly, every morning at these resorts you can watch the German tourists come to the pool to place their towels on the best beach chairs. They leave their towels there all morning, staking their claim to these chairs before the crowds of swimmers and sunbathers arrive. I've always found this picture so amusing, but then I suppose Europe is a crowded place. Europeans are probably used to having to claim parking spots and picnic tables and beach chairs before one of the many other people around claims them.

Here in Australia it's rare to find such crowds. If you're on a beach in Australia and you can look down the beach and see one other person, Australians would call that a "crowded" beach. Australia is a continent of wide open spaces. That's why it felt so strange to be packed in an anchorage with other boats.

The only people visible were a couple of guys two boats down, who were standing on their deck. They were slender and were wearing short swim trunks. One was holding a coffee mug. *They must be European,* I thought to myself. *If they were Australian, they would already be drinking beer at this hour.*

Something irritated me about being moored so close to other people. Prior to this, we had had every anchorage to ourselves. It had been a blissful sense of isolation. At Stonehaven, being so close to other people, we would have to keep our music at a respectful volume. We wouldn't be able to skinny dip off the stern every morning. I wouldn't be able to write to only the sound of the birds in the background. I looked over at the two European guys again, and I noticed they were looking back at me with the same curiosity. They must have also been wondering what kind of person I was. And what earned me the right to be here in this crowded anchorage? I began to wonder how long any of us could possibly hold onto this imaginary territory that we had claimed for ourselves.

II.

We like to have virgin stretches of nature to ourselves. We like to think that we are the only ones with tickets to a private screening of nature's majesty. When we have such intimate encounters with nature, we get a rush of discovery. Like the first explorers, we feel like we are pushing the boundaries of the human range.

Millions of years ago, the Yucatán peninsula between the Gulf of Mexico and the Caribbean Sea was an enormous coral reef. Today rainwater still filters through its porous limestone bedrock and carves out freshwater pockets below. If one of these cavities becomes large enough and the surface is weak, the ceiling will collapse on itself, forming a deep, freshwater-filled sinkhole in the ground, known in Mexico as a *cenote*.[1] It was in a *cenote* that I first experienced the rush of discovery.

There are thousands of *cenotes* across Yucatán, and 900 of them form a ring around the colonial city of Mérida. This is because Mérida sits at the center of an impact crater from an asteroid that struck the peninsula 66 mya. The asteroid weakened the sediment and created concentric rings of *cenotes* around Mérida and across the peninsula.[1]

Cenotes are a popular attraction for tourists and local swimmers alike. They are refreshing private pools, always a cool 24–25°C, hidden deep within the jungle. *Cenotes* also have an otherworldly appeal. The roots of surrounding trees hang down into the dark sinkhole, and the sunlight comes through in fractured rays. It's echoey on the water's surface, making you feel like you're inside a temple.

The real sanctity of a *cenote* is hidden from most swimmers. Lying below the water's surface of the *cenote* is a cave, and this is where the mystery begins. Often there are whole systems of underwater rooms, boasting elegant stalactites and stalagmites and chimneys. Our dive instructor, Fernando, took us to a handful of remote *cenotes* outside the city. Their entrances were narrow, deep, and difficult to access—so for a few brief hours, we had them all to ourselves.

At first sight, navigating through the dark underwater passages of a *cenote* with only a flashlight to illuminate my way was terrifying. After I took some slow breaths to halt my panic, however, I began to feel a thrill unlike anything I had ever experienced. Apart from the flashlight paths of my two dive buddies, the only sign of life was the sound of my own bubbles as I exhaled through my regulator. The bubbles floated up and became trapped under the ceiling of the cavern, creating what looked like polished magnetite gemstones all above me.

Exploring underwater caverns is probably the most explorer-y I will get in this lifetime, and I'm okay with that. I don't have a burning desire to see the deep ocean floor or go to space, for example, like my husband does. But having touched the dark waters of Mérida's *cenotes*, I now understand why he gets such an adrenaline rush from exploration. When you're exploring somewhere on your own, you feel as if you're the very first human to ever go there.

Of course, it only takes the sight of a human skull on the bottom of the *cenote* to remind us that we are not the first to ever go there.[2] Between 1,400 and 800 years ago, there was a thriving Mayan population on the Yucatán peninsula, centered around the city of Chichén Itzá. Chichén Itzá grew to prominence during a population boom that coincided with a period of increased rainfall.[3] The Mayan inhabitants of Chichén Itzá were some of the world's first great astronomers, devising a 365 day calendar and building an astronomical observatory with platforms and doorways that align with the cycles of the Sun, Venus, and the stars.[4]

Surrounding Chichén Itzá are multiple *cenotes*—four are visible, and the others are covered by forest. Although the *cenotes* make for interesting features today, they had much more significance for the people of Chichén Itzá some 1,400–800 years ago. In the Mayan language, Chichén Itzá means "at the mouth of the well of Itzá." Itzá referred to the people inhabiting the region, possibly stemming from the Mayan words *itz*, "sorcerer," and *há*, "water."[5] There were few lakes or rivers on the Yucatán peninsula, so it was at the "mouth of the well" where these water sorcerers established themselves, building one of the largest Mayan cities.

Apart from being a source of water, the *cenotes* held spiritual significance for the inhabitants of Chichén Itzá. They cultivated corn, beans, squash, tomatoes, and chiles for consumption and cotton and cacao for trade,[4] and for all of these crops, they relied on rain.[3] They believed that the Mayan rain god, Chac, lived in the depths of the *cenotes*.[6] They also considered the *cenotes* to be the entrance to *Xibalba*, the underworld.[1] The spirits were thought to flow through the underground rivers connecting the *cenotes*. Hence, to the Mayan people, the *cenotes* were sacred sites—places where they would perform offerings and rituals to please the gods.[1]

At the bottom of a few *cenotes* near Chichén Itzá, archaeologists have discovered jade and gold jewelry, pottery, and incense deposited next to human remains. The Maya might have used these *cenotes* as sacred burial grounds so that their dead could rest at the entrance to the underworld.[1] Some of the human remains in the Sagrado Cenote, or Sacred Cenote, have wound marks on them consistent with human sacrifice. According to Spanish records, the Mayan inhabitants of Chichén Itzá used to throw young women into *cenotes* live, as sacrifices to the Mayan rain god, Chac.[6]

The Maya are often depicted as peace-loving intellectuals in contrast to the blood-thirsty Aztecs,[7] but what is depicted on the brilliantly colored murals in the Upper Temple of the Jaguars and the Temple of Warriors at Chichén Itzá is the commonality of war among the Maya. These murals show a wide array of battle scenes, including entire villages being seized by armies. Along with Mayan hieroglyphic inscriptions, the murals suggest that warfare was commonplace among the various Mayan political groups throughout the Yucatán peninsula.

The resource-rich land surrounding Chichén Itzá would have been prime territory to control. Both the agricultural plains surrounding the city and the nearby marine resources to the north would have been critical to sustaining Chichén Itzá's population and building the city's wealth. Fish and shellfish were a predictable resource that could be harvested using nets, wicker baskets, and hooks and lines. The shells were highly sought after as prestige and ritualistic items. Salt was also important for preserving food. By controlling the Gulf Coast, the people of Chichén Itzá would have controlled shallow saline pools that could be mined for salt.[4] It's possible that the people of Chichén Itzá relied on warfare to expand their territory and gain access to these valuable resources.

The first Spanish to set foot in Chichén Itzá after landing in Yucatán in 1517 were obviously not the first people to recognize the value of the territory nor the first to practice warfare to control it. Pre-Hispanic fortification systems in the cities of Tikal, Aguacatal, Becán, and Muralla de León indicate that various Mayan groups thought there was a need to protect their cities from each other.[8] Similarly, once the Spanish did arrive, the Maya did not just lay down and hand over their territory. It took three conquest attempts of the Yucatán region from 1527 to 1548 before the Maya would finally relinquish their territory.[9]

Our territory is key to our survival. Like many other animals, we use our aggressive behavior to harm, damage, or destroy others if something valuable to us is at stake.[10] What the *cenotes* of Mexico and the civilization at Chichén Itzá demonstrate is that territory has always been valuable to us, both for sustenance and meaning, and that such characteristics make territory worth fighting for.

In November 2017, explorers from the Great Maya Aquifer project came across two underground passages that appeared to lead to a *cenote* hidden directly beneath the largest pyramid at Chichén Itzá. They explored these passages in the hope that they would find the hidden *cenote*, but as they got closer, they discovered that the Maya had intentionally sealed off the entrances with piles of stone. One of the underwater archaeologists on the team, Guillermo de Anda, believes that the Maya sealed the *cenote* because they considered it a portal to the sacred center of the world.[11] Eventually De Anda and his team of explorers came across an entire system of *cenotes* under Chichén Itzá, with a trove of over 150 ritual objects that had been untouched for thousands of years. De Anda said about the discovery, "I couldn't speak, I started to cry. I've analyzed human remains in [Chichén Itzá's] Sacred Cenote, but nothing compares to the sensation I had entering, alone, for the first time in that cave."[12]

But of course, De Anda was not alone. Other humans had been in the *cenote* before, depositing their precious items, burying their dead, communing with their gods, and making this special well of nature their own. By sealing the *cenote*'s entrance, they were still protecting their territory, even thousands of years after they were gone.

III.

My neighbors recently told me about this hotel in Canberra where you share a room with a bear. The hotel is inside the National Zoo and Aquarium, and each room backs onto the exhibit of an animal at the zoo. In the bear's case, you spend the night in a bungalow, and on the other side of a glass wall is a sun bear. You can even see the bear from your bathtub. You get a private window into the bear's life. It's just you and the bear.

Bears would actually make polite roommates. They are solitary animals, minding their own business most of the time. They're also good at letting other animals know where they are. Recently, male brown bears (*Ursus arctos*) were found to deposit scents as they walk, from the pedal glands on the pads of their feet.[13] This may be their way of letting other bears in the area know, "Hey, I'm here."

Although bears like to have their own space, they don't often fight over territory.[13] Brown bears live in forests, mountain areas, tundra, and semi-desert areas of Europe, Asia, and the northwestern part of North America, with home ranges spanning hundreds of square kilometers.[14,

[15] Within these ranges, they give each other plenty of distance. Their social interactions are limited to brief male–female relationships during breeding season and the 1 to 3 years that cubs stay with their mothers before dispersing.[16] In general, brown bears will go to great lengths to leave each other alone. However, as with humans, when a valuable resource is at stake, they are more likely to come into conflict.

In the early 1970s, ethologists Allan Egbert and Allen Stokes were curious about how brown bears act around each other when they're feeding on highly concentrated resources. Brown bears are normally solitary, but they have been known to gather on occasion in loose groups to feed on clumped resources, like garbage dumps and salmon streams. Egbert and Stokes set up camp in Alaska at McNeil Falls. These waterfalls act as a natural barrier to the upstream migration of salmon, trapping and aggregating the fish at the bottom. Observers report that every summer during peak salmon migration, 30 or more brown bears can be present at once at the bottom of the falls. Egbert and Stokes decided to watch the brown bears of McNeil Falls for the 40 days of their summer fishing season.[16]

Salmon began to appear in the first week of July. The first bears to arrive were those who had been there in previous years. They were the expert fishers, and this was their well-trodden spot. During the first days of the fishing season, encounters between the bears were rare. The bears were wary and avoided each other, and often two bears would run away from each other simultaneously.

Over the following weeks, more and more bears arrived at the falls. Bears came from all over the surrounding area to get their chance at harvesting the trapped salmon. As the feeding spot became more popular, interactions between bears became inevitable.

The bears did not erupt into fights straight away. Instead, Egbert and Stokes noticed that as one bear began to encroach on another bear's space, the bears would initiate a defined sequence of behaviors, with each behavior a bit more aggressive than the last. If the intruding bear approached within 4 meters, the defending bear would lower his head below the horizontal line of his body and lay his ears back flat against his head. He would then slowly open and close his mouth while letting out a low, monotone roar. This interaction was as if to warn, "This is my space. Don't you dare come closer."

If, despite this warning, the intruding bear did come closer, the defending bear would escalate his aggression. Egbert and Stokes observed that within close range, about 20 centimeters, the bears would face each other and raise their heads high. Commonly, both bears would extend their heads diagonally upwards, and they would hold their mouths continuously open, giving the impression that they were about to interlock jaws. Their weight would shift to their hindquarters, presumably to free their front limbs for striking. From this position, the bears would make sudden head movements as they let out a continuous, loud roar.[16]

If, after this, one bear still did not forfeit, the next act of aggression would be a hard, fast rush at the opponent. With the head held low and the eyes fixed on the opponent, the charging bear would let out a low growl which would slowly increase in volume to a loud roar as the bear charged towards the opponent. In some cases, the charge was terminated after only three or four strides if the receiving bear began to flee.[16]

If charges did not serve to dissuade one of the bears, then and only then would actual violence ensue. Violent contact involved one bear striking the other bear with one or both forepaws, usually aiming for the chest or shoulder. The strikes were coupled with biting directed at the head and neck. Contact of this type was exceptionally rare, however. Egbert and Stokes noted that it was far more common for one bear to flee at the other bear's approach.[16]

The sequence of behaviors that Egbert and Stokes observed in the brown bears at McNeil Falls is what is known as ritualized fighting.[17] Ritualized fighting is a form of social signaling, a way of letting other animals know to stay back or they will be attacked. It would be too costly in terms of energy and risk of injury for brown bears to attack each other indiscriminately. Instead, brown bears use ritualized fighting to defend their territories and the resources that are important to them. It is their way of intimidating intruders without actually having to engage in fights.

Ritualized fighting was not surprising to Egbert and Stokes. They had seen it in bears before. What was surprising, however, was how the intensity of the brown bears' aggression seemed to depend on the abundance of salmon at the time. From the summer of 1972 to the summer of 1973, the abundance of salmon at McNeil Falls nearly doubled, while the average number of aggressive encounters between bears halved. Over the summer of 1972, each bear caught an average of 1.04 salmon per hour, while in 1973, each bear caught 2.06 salmon per hour. When salmon were scarcer in the summer of 1972, charges between bears were twice as common, and violent contact three times more frequent, than in 1973, when the salmon were more abundant. This relationship between salmon abundance and aggression held within each salmon season as well. Towards the end of the 1973 summer, striking and biting increased as the average number of salmon caught per hour decreased from 2.1 to 1.0. When salmon increased again in days 36 to 40, the fighting sharply declined.[16] The salmon were acting like a dial for the bears' aggression. What was going on?

Almost all vertebrates show aggression, but only from time to time.[18] In animals, aggression is more likely to occur when the right conditions in the brain and the environment align. Neuroscientists have identified three principal circuits in the mammalian brain that play a role in aggression, and the question of which circuit is activated depends on the environmental context at the time.

The first brain circuit of aggression is associated with predatory attack, which is when animals hunt and kill other animals for food.[18] Unlike other types of aggression, a predatory attack is "calm"—it is methodical and focused.[19] Because it is motivated by appetite rather than anger or frustration, predatory attack is thought to be more closely related to the brain circuits involved with seeking food.[18]

Another neurologically distinct type of aggression is inter-male aggression, which is when males fight over females in a form of social competition. This type of aggression happens when the mature males of certain species become territorial of females during the breeding season. They will ward off other males, often in grandiose displays of aggression. For instance, male Galápagos sea lions assume certain aggressive postures during breeding season as a way of telling other males to stay back. In these species, males are generally larger than females, and they experience a hormonal spike in testosterone during the breeding season, which adds to their aggressive displays.[18] After the breeding season, aggression in males declines again.

In humans, predatory attack in the traditional sense is rarely witnessed, although the same system may be involved in what we often consider to be "assertive" behaviors, like asking your boss for a raise, deciding to clean out your closet, or sitting down to write a book.[19] Testosterone-fueled competition between males in the presence of an attractive female, which starts with "showing off" and verbal exchanges and sometimes culminates in a physical brawl, is another type of aggression that you can witness every Saturday night at nearly every pub in the world. But neither assertiveness nor showing off for girls can explain what was bringing the brown bears at the salmon stream into physical conflict or what motivated the people of Chichén Itzá to invade their surrounding villages, or the Spanish to later invade them. For that, animals rely on a third neurocircuit, namely rage.

Rage, or defensive rage as it is more commonly called by behavioral neuroscientists,[20] is what happens in your body when you don't get what you want or need for survival. Rage is mobilized when an animal is frustrated, irritated, or threatened. It involves a rise in blood pressure, focused attention, and invigorated muscles, preparing the animal to either remove an obstacle or subdue an attacker.[19] Rage can be elicited from animals through electrical stimulation of areas of the brain from which instinctual impulses arise, including the corticomedial amygdala, the rostro-caudal extent of the medial hypothalamus, and the dorsolateral region of the midbrain periaqueductal gray.[19] The result of activating the rage circuitry is an automatic, whole-body response in an animal, producing behavior that often includes "marked sympathetic signs, retraction of the ears, baring of the teeth, growling, hissing, piloerection, arching of back, unsheathing of the claws, salivation, and an explosive attack."[21]

Importantly, though, the automatic impulses involved in rage are modulated by other emotional centers in the brain and higher cognitive processes. This means that, even if aggressive tendencies are hardwired or instinctual, cognition will ultimately put a check on when aggression is expressed, as well as the intensity and duration of the aggression.[21] This may explain why bears are aggressive but are also extremely measured in their aggression, rarely letting it escalate to violence.

Brown bears may live far away from each other during most of the year, but when the salmon are concentrated in one spot, they know how to exploit that resource while minimizing the likelihood of full-blown conflict. Egbert and Stokes noted:

> *While there was no group integration and coordination typical of social carnivores, and while individual relationships were flexible, the sum of these factors resulted in formation of a social organization that was relatively stable...salmon abundance determined in large part the degree of social stability. A decline in salmon numbers was reflected by an immediate increase in intolerance among the bears.*[16]

Social stability built on tolerance is the most common way of life for brown bears. But when bears are crowded around a concentrated resource, and that resource on which they depend becomes scarce, violent conflict between bears is more likely to erupt. Sound familiar?

IV.

Humans have essentially the same aggressive neurocircuitry in our brains as other mammals do, yet there is nothing innate within the mammalian brain that can fully explain war.[22] Bears may attack each other every now and then, but they do not form coalitions and start wars. Even our coalition-forming relatives, chimpanzees, who sometimes carry out inter-group attacks, do not decimate their own species to the same extent that modern humans do. So, when and why did humans start waging war?

Popular theories about our warrior nature have relied heavily on comparisons with chimpanzees and studies of modern hunter-gatherer groups. These accounts would lead us to assume that our ancestors have always engaged in war and that humans are predisposed to violence. However, if we consider only evidence from the archaeological record (rock art and human skeletal remains), it appears that warfare in prehistorical times was rather limited.[23] There are isolated examples of interpersonal violence dating back to the Paleolithic, but these do not confirm the occurrence of war.[24] More conclusive evidence of war does not appear until the Mesolithic, and war did not become widespread until the Neolithic, coinciding with the advent of agriculture and sedentary cultures.[24] These archaeological findings would suggest that warfare is a relatively recent phenomenon.

The cemetery of Jebel Sahaba in the Nile Valley is one of the most emblematic examples of widespread violence among early humans. It was discovered in 1964 by archaeologist Fred Wendorf, and remains from the site were analyzed shortly thereafter. The researchers found cut marks consistent with interpersonal violence on at least half of the skeletons, and some skeletons had arrows embedded directly in the bone. The arrowheads appeared to have originated from another culture, the Qadan people. These findings led the research team to believe they had stumbled upon evidence of a full-blown war between the Qadan and Jebel Sahaba people some 18,600–13,400 years ago.[25]

Until recently, it was unclear whether the remains at Jebel Sahaba were the result of a single war or the accumulation of multiple episodes of interpersonal violence. In 2021 a team of French archaeologists visited the British Museum, where the Jebel Sahaba remains are now housed, to take a second look. The team re-analyzed the age and sex of 61 individuals, checking their bones under a digital microscope for lesions and projectile impact marks. They found lesions on 41 of the individuals. Of those individuals who had a lesion, most individuals showed signs of trauma (92.7%, n = 38), and the majority had clear impact marks from projectile weapons (61.0%, n = 25). Interestingly, some individuals appeared to have been injured multiple times, as indicated by previous wounds that had healed. This co-occurrence of healed and unhealed lesions on the Jebel Sahaba skeletons strongly supports the notion that, rather than a single war, there were multiple episodes of interpersonal violence and perhaps warfare between Nile Valley groups at the end of the Late Pleistocene.[25]

It's possible that competition over highly concentrated resources precipitated the violence at Jebel Sahaba. In the second half of the late Pleistocene, the region was extremely arid, but around 14,000 years ago, Lake Victoria suddenly overflowed into the Nile, establishing a pattern of regular flooding.[26] Because of this regular flooding, plants and animals became abundant and concentrated in time and place around the floodplain of northern Egypt and southern Sudan. These conditions meant that human populations could thrive in the Nile Valley and eventually allowed for the establishment of the Egyptian civilization. Conflict might have emerged in the Nile Valley as populations warred over highly concentrated food sources. At least half of the lesions in the skeletal remains from Jebel Sahaba were from projectiles, leading the scientists to believe the injuries were the result of inter-group attacks, rather than conflicts within groups or families. Some of these injuries had obviously healed, suggesting that an individual would have lived through multiple armed conflicts during his or her lifetime.[25]

Highly concentrated resources alone are not enough to guarantee war.[24] At the same time that the people at Jebel Sahaba were battling over resource-rich areas in the Nile Valley, the Natufian people of the Levant were exploiting highly concentrated resources without resorting to war. Like the Nile Valley, the Levant (which includes present-day Syria, Lebanon, Jordan, Israel, Palestine, and most of Turkey) was rich in resources at the time of its occupation by *Homo sapiens* and Neandertals.[24] However, evidence of warfare from this region is extremely scarce. The oldest case of interpersonal violence, a Neandertal male dating back 51,000–47,000 years, was found in the Levant with a partially healed injury in the rib from a weapon, but there is no other clear evidence of other violence in the region.[24] Could it be that the pre-conditions for war were missing from the Levant, or alternatively, that they had solid foundations for peace?

For groups in the same region to live together peacefully, they must generally form social ties that connect their groups. This is achieved, for example, through cross-cultural marriages, where the ethnic separation of groups becomes fuzzier. Peace can also be the result of mutual interdependence and cooperative effort, such as when groups rely on each other to trade essential resources. Strong institutions or established processes to resolve conflicts are also a necessary precursor for peace, because even if you're intermarried and exchange goods, disagreements are bound to arise.[24]

Other social animals have their own established processes for resolving conflicts. Egbert and Stokes noticed that brown bears engage in a series of ritualistic aggressive behaviors as a form of social signaling to prevent full-blown attacks on each other. Similarly, humans in the Levant might have engaged in some form of social signaling to prevent intergroup violence. An abundance of mace heads, which were likely used as symbols of authority, have been found in the Levant, but there isn't any evidence of deaths from mace heads.[24] It's possible that the mace heads were used by the people of the Levant as a form of ritualistic or symbolic aggression, which might have prevented the actual occurrence of war.

The Levant continued for thousands of years in a seemingly peaceful existence, and then almost overnight, war arrived. From around 3300 B.C.E., major settlements all over the region began to put up walls. The destruction of walled cities became common in central Jordan, indicating an abrupt transition to war.[24] At the same time, many goods from Egypt began to appear in the Levant. What was happening during this sudden transition to war?

From 3300 B.C.E., the Egyptian empire was becoming more powerful. To support their growing civilization, the Egyptians needed resources and labor from their neighbors. If you try to force your own people into labor, you can expect political backlash. But if you conquer your neighbors, you can take slaves back with you, isolating them from their home communities and forcing them to rely on you. It appears that the Egyptian empire did just that. In imperialist fashion, they extended their influence east and set up administrative centers in some of the previously peaceful settlements of the Levant. With their invasion, they brought their culture of warfare. This instigated a social change in the Levant that set the region on a dangerous new trajectory.[24]

It is clear that although the capacity for aggression exists in humans, as in brown bears, full-blown violence among early humans was often avoided. Many peacekeeping mechanisms still exist in human societies because peaceful coexistence is more beneficial to all of us than the injury and death toll of war.[27] Warfare requires a massive societal effort to break through these natural checks on aggression. As anthropologist Agustín Fuentes noted, "Groups of monkeys may fight over food sources, ant colonies fight over space, and many other types of animal groups engage in conflicts, but none of them are planned, organized, and lethal with regularity."[28]

Human war, unlike other forms of conflict among animals, requires cultural mobilization. It requires a belief that the enemy is a fundamentally separate entity, or an **Illusion of Separateness**. War requires ideologies of separateness that are so powerful that they can overcome the cooperative tendencies and peacekeeping mechanisms built into human societies. Perhaps early humans did keep peace within their societies, but at some stage, they began to see themselves as separate from those belonging to other groups. Could it be that this **Illusion of Separateness** formed the earliest foundations of war?

War requires the mobilization of entire populations to take up arms and fight each other, which requires a high degree of within-group cooperation and a deep-seated, almost fanatical joint intentionality. And what could foster more intense feelings than the fear of loss? Like brown bears, humans are great at keeping our aggression in check until a resource of vital significance comes under threat. Then, fearing its loss, we break decorum and bare our teeth.

CHAPTER 9 REFERENCES

Free Resources

This 2021 webinar with anthropologist Marc Kissel, which was hosted by the Smithsonian National Museum of Natural History, provides a good overview of the debate around the evolution of human warfare and peace: https://naturalhistory.si.edu/education/teaching-resources/social-studies/webinar-evolution-human-warfare-and-peacefare

Further Reading

Fry, D. P. (2007). *Beyond war: The human potential for peace.* Oxford University Press. https://doi.org/10.1093/oso/9780195309485.001.0001

Fry, D. P. (2013). *War, peace, and human nature: The convergence of evolutionary and cultural views.* Oxford University Press. https://doi.org/10.1093/acprof:oso/9780199858996.001.0001

References

1. *What are cenotes?* (2023). Cenote Finder. https://cenotefinder.com/what-are-cenotes/

2. González, A. H., Terrazas, A., Stinnesbeck, W., Benavente, M. E., Avilés, J., Rojas, C., Padilla, J. M., Velásquez, A., Acevez, E., & Frey, E. (2014). The first human settlers on the Yucatan Peninsula: Evidence from drowned caves in the state of Quintana Roo (South Mexico). In *Paleoamerican Odyssey* (pp. 323–337). Texas A&M University Press. https://www.tamupress.com/book/9781623491925/paleoamerican-odyssey/

3. Hoggarth, J. A., Breitenbach, S. F. M., Culleton, B. J., Ebert, C. E., Masson, M. A., & Kennett, D. J. (2016). The political collapse of Chichén Itzá in climatic and cultural context. *Global and Planetary Change, 138,* 25–42. https://doi.org/10.1016/j.gloplacha.2015.12.007

4. Coggins, C. C., & Shane, O. C. (2014). *Cenote of sacrifice.* University of Texas Press. https://utpress.utexas.edu/9781477302736/

5. History.com Editors. (2023). *Chichen Itza.* A&E Television Networks. https://www.history.com/topics/ancient-americas/chichen-itza

6. *Chichén* Itzá. (2010, November 16). National Geographic. https://www.nationalgeographic.com/travel/world-heritage/article/chichen-itza

7. Jones, L. (1997). Conquests of the imagination: Maya-Mexican polarity and the story of Chichén Itzá. *American Anthropologist, 99*(2), 275–290. https://doi.org/10.1525/aa.1997.99.2.275

8. Ambrosino, J. N., Ardren, T., & Stanton, T. W. (2003). The history of warfare at Yaxuná. In *Ancient Mesoamerican warfare* (pp. 109–123). AltaMira Press. https://rowman.com/ISBN/9780759102828/Ancient-Mesoamerican-Warfare

9. Chamberlain, R. S. (1948). *The conquest and colonization of Yucatan, 1517-1550.* Carnegie Institution of Washington. https://hdl.handle.net/2027/mdp.39015014584406

10. Moyer, K. E. (1976). *The psychobiology of aggression*. Harper & Row. http://hdl.handle.net/10822/545796

11. Tutton, M. (2018, February 2). *Hidden tunnel could lead to Mayan 'entrance to the underworld.'* CNN. https://edition.cnn.com/2018/02/02/world/chichen-itza-maya-tunnel-cenote/index.html

12. *Maya ritual cave 'untouched' for 1,000 years stuns archaeologists*. (2019, March 4). National Geographic. https://www.nationalgeographic.com/culture/article/maya-ritual-balamku-cave-stuns-archaeologists

13. Sergiel, A., Naves, J., Kujawski, P., Maślak, R., Serwa, E., Ramos, D., Fernández-Gil, A., Revilla, E., Zwijacz-Kozica, T., Zięba, F., Painer, J., & Selva, N. (2017). Histological, chemical and behavioural evidence of pedal communication in brown bears. *Scientific Reports*, *7*(1), 1052. https://doi.org/10.1038/s41598-017-01136-1

14. The Editors of Encyclopaedia Britannica. (2024). Brown bear: Mammal, *Ursus arctos*. In *Encyclopaedia Britannica*. https://www.britannica.com/animal/brown-bear

15. Dahle, B., & Swenson, J. E. (2003). Seasonal range size in relation to reproductive strategies in brown bears *Ursus arctos*. *Journal of Animal Ecology*, *72*(4), 660–667. https://doi.org/10.1046/j.1365-2656.2003.00737.x

16. Egbert, A. L., Stokes, A. W., & Egbert, A. L. (1976). The social behaviour of brown bears on an Alaskan salmon stream. In *A Selection of Papers from the Third International Conference on Bear Research and Management, Binghamton, New York, USA, and Moscow, U.S.S.R., June 1974* (Vol. 3, pp. 41–56). International Association for Bear Research and Management. https://doi.org/10.2307/3872753

17. Stirling, I., & Derocher, A. E. (1990). Factors affecting the evolution and behavioral ecology of the modern bears. *Bears: Their Biology and Management. A Selection of Papers from the Eighth International Conference on Bear Research and Management, Victoria, British Columbia, Canada, February 1989*, *8*, 189–204. https://doi.org/10.2307/3872919

18. Panksepp, J. (1998). *Affective neuroscience: The foundations of human and animal emotions*. Oxford University Press. https://academic.oup.com/book/53534

19. Panksepp, J., & Zellner, M. R. (2004). Towards a neurobiologically based unified theory of aggression. *Revue Internationale de Psychologie Sociale 2004 N° 2*, *17*, 37–62. https://mzellner.com/page1/files/2004%20Panksepp%20Zellner.pdf

20. Siegel, A., & Victoroff, J. (2009). Understanding human aggression: New insights from neuroscience. *International Journal of Law and Psychiatry*, *32*(4), 209–215. https://doi.org/10.1016/j.ijlp.2009.06.001

21. Bhatt, S., Gregg, T. R., & Siegel, A. (2003). NK1 receptors in the medial hypothalamus potentiate defensive rage behavior elicited from the midbrain periaqueductal gray of the cat. *Brain Research*, *966*(1), 54–64. https://doi.org/10.1016/S0006-8993(02)04189-6

22. Ferguson, B. R. (2008). Ten points on war. *Social Analysis*, *52*(2). https://doi.org/10.3167/sa.2008.520203

23. Haas, J., & Piscitelli, M. (2013). The prehistory of warfare: Misled by ethnography. In D. P. Fry (Ed.), *War, peace, and human nature: The convergence of evolutionary and cultural views* (pp. 168–190). Oxford University Press. https://doi.org/https://doi.org/10.1093/acprof:oso/9780199858996.001.0001

24. Ferguson, B. R. (2013). The prehistory of war and peace in Europe and the Near East. In D. P. Fry (Ed.), *War, peace, and human nature: The convergence of evolutionary and cultural views* (pp. 191–240). Oxford University Press. https://doi.org/10.1093/acprof:oso/9780199858996.001.0001

25. Crevecoeur, I., Dias-Meirinho, M.-H., Zazzo, A., Antoine, D., & Bon, F. (2021). New insights on interpersonal violence in the Late Pleistocene based on the Nile valley cemetery of Jebel Sahaba. *Scientific Reports, 11*(1), 9991. https://doi.org/10.1038/s41598-021-89386-y

26. Williams, M., Talbot, M., Aharon, P., Abdl Salaam, Y., Williams, F., & Inge Brendeland, K. (2006). Abrupt return of the summer monsoon 15,000 years ago: New supporting evidence from the lower White Nile valley and Lake Albert. *Quaternary Science Reviews, 25*(19–20), 2651–2665. https://doi.org/10.1016/j.quascirev.2005.07.019

27. Fry, D. P. (2007). *Beyond war: The human potential for peace*. Oxford University Press. https://doi.org/10.1093/oso/9780195309485.001.0001

28. Fuentes, A. (2013). Cooperation, conflict, and niche construction in the genus *Homo*. In D. P. Fry (Ed.), *War, peace, and human nature: The convergence of evolutionary and cultural views* (pp. 78–94). Oxford University Press. https://doi.org/10.1093/acprof:oso/9780199858996.001.0001

CHAPTER 10

Rebel Rebel: Individualism

I.

It was a couple hours after sunrise at Stonehaven, and the boat had hardly roused. From the cockpit I observed the water. It was a deep green, having swallowed all the reflections of the forested shore. It reminded me of Stockton Lake, Missouri, where my parents used to take us sailing when we were kids. Naturally, I love the exotic aquamarine of the Bahamas or the Sea of Cortez, but a glossy, veridian water, which in its depths appears to house the core of the Earth, always feels like home.

To my right, at the shallowest end of the cove, was an old fishing boat with a peppermint cabin and a dirty white hull. My eyes followed the lines of the boat down to the water's surface, where there was a mint and white replica of it, its outline distorted like molten lava.

Boats are like people. They're unique individuals with their own stories. The shape of their lines, the colors and font used on the hull, the growth of soft corals around the waterline, the stylistic choices of the boom cover and bimini, the gadgets—all of these reflect the history of their designer and owners. In a crowded anchorage you get to gaze at a whole spectrum of oddities and wonder, what kind of person would want such a boat?

My ringtone startled me. It was the first time I had heard it in days. The blaring melody jolted me back to my former life—running to meetings on the other side of campus, phone in hand, frantically looking up the staff profile photo of whichever professor I was about to meet. I hadn't known any stress like that since we set sail 6 days earlier. I followed the ringtone down into the saloon, but my phone was nowhere in sight. I looked under the pillows on the sofa to no avail, as it continued to ring. Then I opened the nav table, and there it was, the screen displaying a number I had never seen before. *It must be someone from the university*, I thought. I pulled together my best professional tone to answer.

"Hello, this is Jill," I chimed.

"Hey Jill, it's Jason," the man responded. I stood up straight. *Why is the charter company calling me? We're not meant to be back until tomorrow.* "We've just received a call from one of the boats near you saying that you're on a broken mooring."

"Oh, really?" I said, mortified. The mooring had seemed fine the day before. It had attached to our bow cleat without any issues.

"You might have noticed when you picked it up yesterday, that it was orange with a big white X painted on it?" he asked.

Crap. It was orange.

"Oh, right," I said. "We didn't see an X." To be honest, I wasn't paying any attention to the mooring ball. It could have had a purple Barney painted on it, for all I knew.

"Well, your neighbor has spotted it and reported it as broken. You're best-off detaching from it and hooking onto another mooring. Otherwise, the boat might drift."

"Oh, okay, of course. We'll move now. Thanks so much for letting us know."

"You should thank your neighbor who reported it," he said. "Lucky there wasn't any wind last night or it could have ended badly."

I felt my cheeks burn. "Yeah, it's a good thing," I responded. "Thanks for the call."

"No worries. Catch ya," he said and hung up.

As soon as the call was over, I collapsed onto the cushion next to the nav table. I pulled my knees into my chest. I couldn't believe someone was spying on us. A wave of adrenaline rushed through my body. It had to have been those pretentious European guys in the short shorts. They would have relished the chance to report our stupidity. Not all of us are born "yachties," after all.

I woke my husband to tell him the news. At first, he looked dumbfounded. Then, the whites of his eyes widened, and his face flushed with color. "I can't believe someone reported us ... why wouldn't they just call?" he said.

I stared at the window. "I know," I said.

He shook his head, his forehead scrunched. "It's good we didn't crash."

"Yeah," I admitted. "Better to know now before the wind picks up."

My husband grabbed the boat hook, and we hurried up the companionway stairs and along the side of the boat to the bow. He reached down with the hook and pulled up the barnacled rope attached to the mooring ball. Sure enough, there was a giant X on it in reflective white tape. Below the X in thick Sharpie, it read, "DO NOT USE." My husband looked back at me. *Well, that was a bit of an oversight.*

Rather than move to another mooring ball, we decided to pack up and leave the anchorage altogether. Suddenly it felt stuffy there.

As we turned the boat around and headed out of the cove, I looked over where I had seen the two European men the previous day. I couldn't see them. *They must be hiding,* I thought. I stared at the windows along their cabin as my right thumbnail dug into the hangnail on my right middle finger. *It has to be them,* I thought, *and now they're not even brave enough to show their faces.*

It's annoying to be called out. Infuriating, actually. But I suppose if no one speaks up when something is amiss, then as the group-minded animals that we are, we might just be headed towards a shipwreck without anyone having the courage to stop us.

II.

At the age of 19, I nearly became Catholic. Nearly, but I couldn't bring myself to go through with it.

It was Easter Sunday. From the back of the church in Coral Gables, Florida, I could barely see the altar, with the pews packed with people dressed to the nines and families standing all along the back and sides of the little church. The priest was cloaked in purple, symbolizing Christ's resurrection. The smell of incense drifted over us like a spell.

My catechism teacher gathered her clutch of adult students around her and reminded us of what would be involved with our initiation. We would walk down the center aisle and up to the altar, where the priest would be waiting with the big book of the Saved. We would write our names in the book, and this would symbolize our official commitment to join the Church. We would be welcomed into the Catholic family of the Saved.

"What about my family in Missouri, who are Methodist?" I remember asking my teacher.

Her penciled-on eyebrows raised, and with a shrug, she said, "Well, we can only say that the people in the book will definitely be saved. We can't say what will happen to everybody else."

And just like that, my entire Protestant family was swept into the depths of fiery hell.

When the right time in the Mass arrived, we walked up the aisle and ascended the altar stairs. The two in front of me, upon approaching the priest, grabbed the pen and scribbled their names in the book, after which the priest blessed them and welcomed them to the Church.

When it was my turn, I stared down at the big book. I saw all the handwritten names on the page and behind them I imagined the patient eyes of my mother and father. I burst into tears. The priest looked uncomfortable. "I'm sorry," I wailed at his wrinkled, pale face, which I could barely see through my flooded eyes. "I'm sorry. I can't." I turned around and made a bee line for the massive wooden doors at the back of the church, never to return.

I spent the next year feeling lost. I had always been someone who was drawn to spirituality, and yet here I was, 19, alone in the vast, hedonistic ocean of Miami without a rudder. I was so sick of religious pomposity that I couldn't bring myself to go to a church of any kind. That's when I remembered my grandfather's account of his visit to a Quaker meeting in St. Louis, Missouri. He said that it was the oddest thing: He entered the room, sat down, and spent the next hour with the Quakers just sitting in silence. There was no preacher, no music. Just silence. Silence sounded good to me after having endured the countless up-down-up-down, credo-reciting liturgies of Catholic Mass. Oddly enough, my grandfather had also said that the Quakers ascribed to no formal creed or dogma. *What unites these Quakers?* I wondered.

When I first stepped foot into the Miami Friends' stone courtyard, tucked behind a thicket of tropical trees and bushes, I found out what unites the Quakers. I was given no book, no pamphlet—just a welcome smile and a handshake from a skinny older man in smudged rectangular glasses and a long white beard, who introduced himself as David. He told me to come in and have a seat anywhere.

The rectangular room was illuminated by the diffuse light of windows that ran all along one side of the building, where a humid breeze was coming across the foliage-shaded louvres. A more intense light poured in through a high triangular window on the back wall, warming the exposed wooden eaves. The room looked like it was set for an orchestra, with metal fold-up chairs arranged in three rows on each wall, all facing the center. But in the center of the chairs was no conductor or podium—just empty space.

The meeting began without announcement. At a certain time, everyone, as if by telepathy, just went quiet. They closed their eyes and slowed their breath. Within 30 seconds, the only sound that could be heard was the oscillating of the standing fan in the corner and the buzz of cicadas outside, superseded occasionally by the sound of a passing car. I soon felt the heaviness of my own breath. Silence descended on my mind like a soft blanket.

Maybe half an hour or 45 minutes later, a voice broke the silence. It was like a chandelier crashing to the floor. I didn't know if I was supposed to open my eyes, so I kept them shut. An older man rambled for a few minutes in a stream of consciousness. His sentences had depth to them, almost like poetry. He finished as abruptly as he had started and sat down. Silence overcame us all again. A few minutes later, a woman spoke, sharing a seemingly unrelated message to the first. People took the opportunity to shift in their chairs. Again the woman stopped speaking suddenly and sat down, and the silence resumed. Waiting for the next person to speak became an exciting game for me, like listening for those last few kernels of popcorn in the microwave bag to pop.

Several minutes after the last message, chairs began to rustle. Quiet chatter arose around me, and I opened my eyes. The people in front of me shook each other's hands, then they turned around to shake my hand. I expected an onslaught of questions as a newcomer, but instead I received smiles and warm, human handshakes.

After the handshakes, people sat back down, and announcements were made by a man who introduced himself as the clerk. Then everyone moved into the next room for refreshments. Voices became louder and more animated as people stood up and launched into longer conversations with each other. I joined the line for coffee next to the kitchen, noting that absolutely nothing in the room was disposable. We all used the same silver spoon for sugar, stirring our coffees and putting it back in the sugar jar. I liked the simplicity of it. It felt like visiting my grandmother's house.

On the table were several plates of healthy snack foods, one of which was a beautiful arrangement of halved figs. I had never eaten a real fig before, only Fig Newton cookies. I grabbed one of the fig halves from the plate and brought it to my mouth as I watched the others chatting, most of them Baby Boomers with graying or fully white hair. By their simple dress and relaxed way of speaking, they seemed to still be living by the mantra of their 1960s cultural revolution—simple, down-to-Earth living, where all you need is peace and love. It was like my grandfather had said, there was no dogma, there was no creed, just individual people who served as the testimony of the Quaker beliefs.

Since their beginnings in England in 1646, I would come to learn, the Quakers, also called the Religious Society of Friends, have been shaped by their strong individual consciences, or what they refer to as the "light within."[1] They believe that each human has direct access to the divine if they can quiet their mind enough and sink down to the seed of the divine within them. Some Quakers identify as Christian; others do not. What unites them all is a central belief that the light within dwells within each individual.

Having direct access to God within oneself may not sound that controversial today, but it was blasphemy during the English Civil Wars, when the Quakers started as a small group of rebels. Throughout the 1640s, one issue that had been raised by Martin Luther and was still being hotly debated was whether regular people had direct access to God or whether they needed to go to through a priest or high religious authority. In the Catholics' view, only a few people could interpret the will of God, including the kings whom God had chosen himself. Meanwhile, Protestants in the same vein as Martin Luther believed that everyone could access God directly through prayer.

The founder of Quakerism, George Fox, traveled throughout the English countryside during the 1640s in search of answers to his spiritual questions. Upon entering a towering steeplehouse in Nottingham, Fox heard the priest say that the Scriptures were the judge by which to try all doctrines, religions, and opinions. According to Fox, "Now the Lord's power was so mighty upon me, and so strong in me, that I could not hold, but was made to cry out and say, 'Oh no, it is not the Scriptures,' and was commanded to tell them that God did not dwell in temples made with hands."[1] Fox was promptly arrested and imprisoned. According to Quaker historian Rosemary Moore, "This was the first of many occasions when Friends were in trouble with the law."[1]

That's a modest statement, which is very ... Quakerly. Throughout the last three centuries, Quakers have been some of humanity's most passionate activists, risking imprisonment for social justice. They were involved in emancipating George Washington's slaves, much to his chagrin. They were key conductors in the Underground Railroad, which concealed escaped slaves from house to house on their journey north to the Free States or Canada.[2] Similarly, just before Nazi Germany invaded Poland, British Quakers arranged for travel papers to be issued to Jewish children in Germany, Czechoslovakia, Austria, and Poland so that they could board Quaker-chaperoned trains to Great Britain and escape Nazi persecution. In what became known as the *Kindertransport*, 10,000 refugee children made the journey out of Nazi-occupied Europe before the war.[3] "Let your life speak," the Quakers would say.

Of course, a religion that revolves around "letting your life speak" doesn't come without its share of controversy. As U.S. involvement in the Vietnam War escalated in the late 1960s, some Quakers became increasingly concerned about the suffering of the people in North Vietnam. They applied for permission to send $6,000 of humanitarian aid to North Vietnam, but the U.S. Government refused. At that point, knowing that proceeding with the aid would put them in violation of the *Trading with the Enemy Act of 1917*, most Quakers decided to abandon the mission. Nevertheless, a small radical group of Quakers, called A Quaker Action Group, decided to persist. In February 1967, A Quaker Action Group chartered a sailing vessel, loaded it with medical supplies, and sailed it into Haiphong harbor in North Vietnam. There they worked with other peace organizations to distribute the medical supplies to the North Vietnamese people. Upon their return to the United States, their passports were confiscated.[4]

Throughout human history, a few brave individuals have gone against the law to take a stand against injustice. Some Quakers obviously felt that their duty to help their brothers and sisters in Vietnam was more important than their American nationality or their cohesion with the larger body of Quakers. Expressing our consciences is a core part of being human. But if expressing one's conscience means breaking away from the main group, what does that mean for the stability of cooperation?

III.

Picture a school of mackerel. The school moves in synchrony, with every individual falling into line. Each turn they make, the entire school turns in unison. They swim this way because it has had advantages for the survival of their species. Swimming in a school or shoal usually affords individual fish with greater protection from predators.[5] The greater the predation risk, the more closely affiliated the fish tend to be.[6] Looking at the mackerel school from the outside, it can be tempting to see the fish as all the same, as interchangeable units within one cohesive cluster. But as environmental conditions or dynamics within the group change, for whatever reason, some individuals may break away.[7]

Sharks are big fish, but fish nonetheless. Although many sharks are solitary predators, roaming the vast stretches of the ocean on their own, they still have complex social lives.[8] Gray nurse sharks (*Carcharias taurus*), also known as sand tiger sharks, are placid-natured and slow-moving sharks who live in temperate subtropical waters all over the world. Here on the east coast of Australia, gray nurse sharks can be found swimming close to shore near sandy gutters, rocky caves, and tropical reefs, making them one of the most common sharks sighted by snorkelers, despite their critically endangered status.[9] They are prone to congregate in groups (fusion), but just like other social fish, sometimes individuals will break away from the group (fission).

To understand the fission–fusion dynamics in gray nurse sharks, oceanographer Danielle Haulsee and colleagues captured 20 adult gray nurse sharks in Delaware Bay one summer. They tagged the sharks with Vemco mobile transceivers before releasing them again. The transceivers picked up and stored coded acoustic pings from animals with compatible transmitters within a few hundred meters. Of these 20 sharks, the research team was able to recapture two males almost a year later. During their annual migration, these two male gray nurse sharks acted as data collectors, recording tens of thousands of transmissions from tagged fish in the Atlantic Ocean.[8]

What did these two boys get up to? It appears that they spent the rest of the summer foraging around Delaware Bay with other gray nurse sharks. There was a large meet-up of sharks in September, with the two males encountering over 100 other gray nurse sharks. This could have been because of the abundance of fish in Delaware Bay at the time. Once October rolled around, the group of gray nurse sharks broke up. The males began their migration south along the east coast of the United States, whereas the females went elsewhere. Also during this fission, the group shifted from being well mixed in terms of body size to a group of almost exclusively similar-sized males. During their migration south, the males encountered many of the same individuals again and again.[8]

Upon reaching the Carolinas in December, many of the migrating sharks began to aggregate again, forming a community bottleneck. The bottleneck was likely due to the narrowing of the continental shelf near Cape Hatteras, North Carolina. There are many shipwrecks around Cape Hatteras, so it could also be that the gray nurse sharks gathered at this spot to take advantage of the abundance of prey fish swimming around the wrecks. At this location, the females also rejoined the males, so it could be that the spot was known by the sharks as a mating area.[8]

Whatever caused the fusion of gray nurse sharks around Cape Hatteras in December, by March the group had split up again. The two data-collecting males appear to have gone off on their own, entering a solitary phase. No one knows why this fission event occurred. Many other groups of animals, including primates and dolphins, have been observed to split up for days at a time in order to reduce the competition for resources, like food, mates, or habitat. But these sharks remained apart for months. By May, it was time for the sharks to migrate north again. As with the southern migration, the boys united and were joined by many other gray nurse sharks, mostly males. Together they made their way back to their summering grounds in Delaware Bay by mid-July.[8]

The movement of the two data-collecting males between Delaware and the Carolinas revealed a complex pattern of fission and fusion in the Atlantic gray nurse shark population. With studies of group dynamics, it's always difficult to know whether external or internal forces are driving these fission and fusion events. In the case of the gray nurse sharks, the underwater topography as well as abundance of fish at certain locations might have brought them together, while nutritional stress or their mating rituals could have forced them apart. Looking more closely, the two data-collecting males did tend to spend more time around certain individuals than others, especially males of their own size. This indicates that the associations among gray nurse sharks are not entirely random.[8] In other words, the sharks may be actively choosing to socialize with certain individuals.

The idea that sharks choose with whom they interact based on their individual preferences may have sounded far-fetched 20 years ago, when such complex behavioral dynamics were associated only with mammals. However, recent studies have confirmed fission–fusion dynamics based on personality or individual preferences in a wide variety of fish. Behavioral ecologist Darren Croft and colleagues studied guppies and found that within their schools, guppies tend to hang out with the same individuals, sorting themselves into "social cliques."[10] Similarly, researchers from the University of São Paulo and Pennsylvania State University have found that rainbow trout placed in a four-chamber experiment showed individual preferences for color. Although all 20 trout preferred blue as a background color, one trout also liked green and another liked yellow. Individual trout differed in the kinds of shelter they preferred as well. Four of the trout preferred no shelters, five preferred to have just 1 shelter, nine trout preferred 2 shelters, and five trout preferred 3 shelters.[11] These findings confirm that even within a single population of fish, not all individuals have the same desires, aversions, and motivations. It's possible that fish, just like humans, choose whom they want to hang out with based on their individual preferences.

Preferences are just one facet of an individual's personality. I prefer the mountains to the beach, but this alone doesn't define me as a person. Apart from mere preferences, like blue or green, an important component of an individual's personality is how that individual responds to new situations. Observations of juvenile lemon sharks (*Negaprion brevirostris*) in the Bahamas have confirmed that some individual sharks are willing to leave the safety of their home environment of mangrove-fringed lagoons and venture out into the deep ocean, whereas many other individuals never venture out.[12] Exploring this notion that some sharks may be "bolder" than others, researchers at the Bimini Biological Field Station Foundation in the Bahamas captured wild juvenile lemon sharks and placed them in a large circular holding pen constructed in shallow, sand-bottom waters. There, one by one, the sharks were subjected to a novel open field test, which involved opening a sliding door to an adjacent pen that none of the sharks had ever seen before. From a nearby tower, the researchers observed the movements of each shark as they explored the new pen. The researchers observed consistent individual differences in the sharks' willingness to explore the new environment. Some sharks moved quickly about the new pen, exploring every corner of it, while others were less exploratory, indicating that they were likely stressed by the new environment.[13]

Why would evolution favor these differences among individuals, if they just cause groups to split apart? According to biologists Max Wolf and Franz Weissing, personality differences among animals determine the behavioral diversity of a population, which can have major consequences for the survival of the species.[14]

I remember a friend in Germany asking my thoughts, while he was chopping onions for spaghetti carbonara, on why there are still so many cultures in the world, if cultural differences just lead to war? I thought about it a bit and answered that if a major ecological disruption were to happen, you would want multiple, diverse solutions for how to respond; otherwise, the whole human race could be wiped out by it. Where there are many individual differences within a group, there is high behavioral diversity, meaning that there are many ways to respond to a change in the environment. I since learned that the traditional Italian way of making spaghetti carbonara does not include onions, but this German friend of mine felt compelled to include onions. In response to his question, cultural differences like these might one day save our species.

Even within smaller populations, like the Bahamian lemon shark population, behavioral diversity can have drastic effects on how stable, resilient, and persistent the population is over time. Behavioral diversity can also affect how the population disperses and forms new populations.[14] In other words, the more eclectic the group of individuals you bring together, the more resilient that group will be to change.

As seen with the gray nurse sharks during their coastal migrations, personality differences may determine how likely a subsection of a group is to fission, splitting from the rest of the group, and how likely individuals are to fuse to form a new group. These individual personalities within any given group are what ultimately determine the extent of the group's cooperation and survival. Simply put, as animals, we are more than a group. We are a mix of individuals, and the nature of that mix matters for the success of our population as a whole.

IV.

Individuality has a biological basis. We can see it in the brain of even the tiniest creatures. If you place genetically identical fruit flies (*Drosophila melanogaster*), reared under the same conditions, in a circular arena with two equally attractive visual landmarks on either side, some individuals will consistently take a straight and narrow path between the two landmarks, while other individuals will take a wide, meandering path. Researchers found that these individual differences in behavior can be attributed to randomly occurring differences during the development of the flies' brains. Despite their identical genetics and rearing, some individuals end up with a greater degree of left–right asymmetry in the neurons projecting to the medulla portion of their optic lobes. These slight differences in brain symmetry mean that individual fruit flies actually see the world differently, which causes them to take different paths when they walk.[15] So even down to the level of a fruit fly, individuality is a product of biology.

In most modern human societies, our individuality matters to us. Human individuals have rights, freedoms, and the ability to vote. If they wish, they can abandon their birthplace and family and travel to the other side of the world to pursue their dreams. The idea that individuals should, within reason, be able to express their individuality only seems natural to us now. But it wasn't always this way.

It is commonly assumed that in prehistoric times, humans lived in collectivist societies, giving preference to the needs of the group over the needs of the individual. Given that collectivism and individualism are cultural constructs, however, it's hard to be certain. For most of hominin evolution, living in groups would have favored our ancestors' survival.

It would have been almost impossible for individuals to survive on their own. But how much importance, if any, individuals were given within early hominin tribes is unknown.

What is clear is that by the Bronze Age, around 3300 B.C.E., the beginnings of individualism in the economic sense were apparent in some major civilizations but not in others. For instance, the Egyptian civilization (3100–332 B.C.E.) thrived in the fertile flood plains along the Nile River, but individualism did not. The Egyptians cultivated barley and wheat and traded them with other regions, but all crop production in Egypt was centralized under the control of the Pharaoh. The Pharaoh owned all the land, distributed all the resources, controlled all the slaves, and made all the decisions. Ordinary people did not have any rights under Egyptian law, so individualism was practically nonexistent.[16]

On the other hand, just 1,300 km east of ancient Egypt, the Mesopotamian civilization (3400–539 B.C.E.) did provide fertile ground for individualism to grow. Along the Tigris and Euphrates rivers, there were two regions known for their natural resources. Southern Mesopotamia provided excellent soil for agriculture, whereas northern Mesopotamia had abundant stone, timber, and bronze. Several Sumerian tribes lived throughout the two regions, and many settlements had grown to the size of city-states. These city-states traded goods with each other, forming a large, loosely connected region. Because of the variable distribution of resources across Mesopotamia, each city-state specialized in the refinement of a particular resource. Specialized craftsmen and women soon emerged. Merchants purchased these specialized goods from the craftsmen and women and took them to other regions to trade. These craft industries often required land and labor, so markets developed in which individuals could buy land and slaves.[16] It is here amid the commerce of Mesopotamia that we see humanity's first written laws about interactions between individual citizens and the first mention of private property.[17]

Further east in the Indus Valley, individualism was beginning to bloom in the religious sphere. There, a transformation from collectivism to individualism in ritual practice began within the Vedic civilization (1500–500 B.C.E.). While earlier Vedic writings emphasized collective ritual (*karman*) and the interdependence of all beings, later Vedic writings refer to new forms of knowledge connected to the "self" (*ātman*). This concept of connection with oneself formed the basis of Eastern religions that encouraged the renunciation of society, such as Buddhism and Jainism.[18]

At the same time that the inhabitants of the Indus Valley were reflecting on knowledge of the self for religious purposes, the ancient Greeks were bringing individualism into the political sphere. "Know thyself" is the first of the Delphic maxims inscribed in the forecourt of the Temple of Apollo at Delphi, reflecting the Greek political philosophy that dates back to at least the 5th century B.C.E. The maxim is attributed to the Seven Sages of Greece, and it was later embraced by Greek philosophers Socrates and Plato. For Socrates, individualism consisted of taking oneself seriously as an object of existential and moral inquiry. He believed that everyone in society should want to know, at the very least, who they are. Socrates's critiques of Athenian democracy in *Republic* give us a glimpse of how the ancient Greeks thought about the individual.[19] Plato insisted that the key to democracy is man's love of freedom, where freedom is a condition in which "a man may say and do what he likes."[20]

During the Middle Ages (476–1350 C.E.) in Europe, the freedom of most individuals was restricted by a handful of elite, who controlled all intellectual, political, and social life. Most Europeans lived in feudal societies, where their labor was all in service to a lord or master. The truth was set and administered by the Church, and total obedience was expected. Although individual "souls" were recognized by the Church, and the concept of individual salvation became popular, the collective still took priority over the individual.[21] It was under the collectivist system of feudalism and the Church's heavy hand, however, that the seeds of individualism planted by the ancient Greeks finally began to sprout.

The Greek ideal of humanism—cultivating human virtue in service of society—would flourish in Florence from the late Middle Ages. After Countess Matilda of Tuscany died in 1115, the people of Florence decided they no longer wished to be under the rule of the March of Tuscany. Instead, they formed an autonomous commune by which to govern themselves democratically. As European demand for woolen cloth and foreign banking grew during the 12th and 13th centuries, the new Republic of Florence grew in size and power. In 1252, Florence introduced Europe's first gold coin, the florin, which became a world monetary standard. Florence soon became a cultural epicenter and one of the most prosperous cities in all of Europe.[22]

Artists, poets, and scientists flocked to Florence during the late 13th and 14th centuries. Their interest in nature, learning, and individualism spread throughout Italian literature and art.[23] In the early 1400s, following the devastation of the Black Death, the Italian humanist Leonardo Bruni became secretary to the papal chancery and later chancellor of Florence.[24] Bruni sparked a revolution in civic discourse that became a major pillar of the Italian Renaissance. During this rebirth of individualism, Bruni established and defended civil liberties and led his fellow citizens away from rule by the elite and towards a new sense of community based on the active virtue of individuals.[25]

In a funeral speech he gave in 1428,[26] Bruni said,

> *Our form of government aims at achieving liberty and equality for each and every citizen. Because it is equal in all respects, it is called a popular government. We tremble before no lord nor are we dominated by the power of a few. All enjoy equal liberty, governed only by law and free from fear of individuals.*[27]

Like the ancient Greeks, the early Italian humanist philosophers and statesmen focused on the cultivation of the individual to benefit society. Initially they translated the Greek Classics and works of natural sciences into Latin, but in the later part of the 15th century, they began translating them into the Florentine vernacular, which would eventually become the Italian language.[28] This switch from Latin to vernacular signaled a shift in cultural beliefs about the average person and his or her ability to understand high literature.

As the humanist ideals of the Renaissance spread to various European cities and principalities governed by the Holy Roman Empire, philosophers and theologians began to seriously question the importance of the individual. Both Johannes Gutenberg's development of the mechanized printing press and the Protestant Reformation placed greater power in the hands of individuals to read and interpret the Christian Bible for themselves. At the same time, the individual was being redefined in political terms. Political philosopher Larry Siedentop has suggested that the first appearance of the words "individual" and "state" in 15th century English and French dictionaries was no coincidence. Individuals were defined by their role in the creation of states—the concept of one was necessary for the other.[21]

Should an individual really be allowed to do whatever he or she likes? Thinkers during the 17th and 18th centuries, or the Age of Enlightenment, reflected on this question at length. Living in a society brings advantages to the individual but also demands sacrifices of personal freedom. Enlightenment thinker Jean-Jacques Rousseau referred to the arrangement between the individual and the group as the "social contract." According to Rousseau, people who live in societies exchange their independence to be part of a larger republic that acts in the interest of the common good. Rousseau suggested that if the government takes too much power from the people, the social contract is broken, and the people have an obligation to rebel.[29] In other words, for human groups the social contract is that fine line between fusion and fission.

Nowhere can we see individualism skirting the line between fission and fusion more than in 18th-century Paris. From the mid-1700s, Paris became the center of philosophical and cultural developments that challenged the traditional order of rule by the elite. On the philosophical side, Voltaire and Rousseau were arguing for a society based on reason, rather than on authority and religious doctrine. And on the cultural side, it was becoming more socially acceptable to express one's individuality by smiling.

It's easy to assume that humans have always used smiling as a way of expressing kindness and good intentions to others, but historian Colin Jones argues that this was not always the case. During the dynastic rule of Louis XIV, even the King of France could not access good dental care, losing almost all of his teeth and one side of his upper jaw to dental disease. Therefore, the French royalty and high society were notoriously tight lipped, not wanting to show the gaping smile that was expected of beastlike peasants. All of this changed during the mid-18th century, when the advent of modern dentistry techniques in France made mouths more palatable. Soon thereafter, smiling emerged as a form of social signaling among the Parisian salon society. Suddenly, Parisians were keen to express their emotions and let others peer into their souls through that tiny window in the mouth, the smile.[30] In doing so, they opened the floodgates for individualism to sweep through France.

For France and many other democratic nations, the fall of the Bastille on July 14, 1789, symbolized the triumph of the individual and the end of collectivist, religiously sanctioned authority. On the one hand, the storming of the Bastille asserted in concrete terms that individuals can and will overthrow a government with which they are dissatisfied. And on the other hand, there was the philosophical assertion that individuals matter. The French Revolution asserted that ordinary people have unique ideas and sentiments, and they should be allowed to express them.[31] With the French Revolution began the modern era of institutionalizing individualism in both social and political life.

At one extreme, humanity can exist in a state of complete fusion, as in ancient Egypt, where people's individual natures are suppressed for the benefit of the group. At the other extreme, humanity can end up in utter fission, as in revolutionary France, where individuals refuse to submit to any kind of external authority. Viewed through the lens of evolution, individualism is a cultural adaptation for protecting behavioral diversity, which is ultimately good for the survival of humanity. The Quakers have demonstrated how individualism can encourage a small group of people to follow their consciences, which can ultimately lead to something positive for society. But with the dawn of the post-modern era, we'll see that there are times when individualism can be taken too far.

CHAPTER 10 REFERENCES

Free Resources

Lukes, S. M. (2024). Individualism. In *Britannica*. https://www.britannica.com/topic/individualism

Further Reading

Jones, C. (2017). *The smile revolution in eighteenth-century Paris*. Oxford University Press. https://global.oup.com/academic/product/the-smile-revolution-9780198715825

Siedentop, L. (2014). *Inventing the individual: The origins of western liberalism*. Harvard University Press. https://www.hup.harvard.edu/books/9780674979888

References

1. Moore, R. (2020). *The light in their consciences: Early Quakers in Britain,1646–1666*. The Pennsylvania State University Press. https://www.psupress.org/books/titles/978-0-271-08589-0.html

2. *Underground railroad*. (n.d.). Quakers in the World. Retrieved March 21, 2024, from https://www.quakersintheworld.org/quakers-in-action/115/Underground-Railroad

3. Britain Yearly Meeting of the Religious Society of Friends (Quakers). (n.d.). *Quakers and Kindertransport*. Quakers in Britain. Retrieved March 21, 2024, from https://www.quaker.org.uk/faith/our-history/kindertransport

4. *Peace witness and relief efforts during the Vietnam War*. (n.d.). Quakers in the World. Retrieved March 21, 2024, from https://www.quakersintheworld.org/quakers-in-action/315/Peace-Witness-and-Relief-Efforts-during-the-Vietnam-War

5. Seghers, B. H. (1974). Schooling behavior in the guppy (*Poecilia reticulata*): An evolutionary response to predation. *Evolution, 28*(3), 486. https://doi.org/10.2307/2407174

6. Kelley, J. L., Morrell, L. J., Inskip, C., Krause, J., & Croft, D. P. (2011). Predation risk shapes social networks in fission-fusion populations. *PLoS ONE, 6*(8), e24280. https://doi.org/10.1371/journal.pone.0024280

7. Croft, D. P., Arrowsmith, B. J., Bielby, J., Skinner, K., White, E., Couzin, I. D., Magurran, A. E., Ramnarine, I., & Krause, J. (2003). Mechanisms underlying shoal composition in the Trinidadian guppy, *Poecilia reticulata*. *Oikos, 100*(3), 429–438. https://doi.org/10.1034/j.1600-0706.2003.12023.x

8. Haulsee, D. E., Fox, D. A., Breece, M. W., Brown, L. M., Kneebone, J., Skomal, G. B., & Oliver, M. J. (2016). Social network analysis reveals potential fission-fusion behavior in a shark. *Scientific Reports, 6*(1), 34087. https://doi.org/10.1038/srep3408

9. Queensland Government. (2022, July 7). *Grey nurse shark-diver interaction research project.* https://www.qld.gov.au/environment/coasts-waterways/marine-parks/projects/grey-nurse-shark

10. Croft, D. P., Krause, J., & James, R. (2004). Social networks in the guppy (*Poecilia reticulata*). *Proceedings of the Royal Society of London. Series B: Biological Sciences, 271*(suppl_6). https://doi.org/10.1098/rsbl.2004.0206

11. Maia, C. M., Ferguson, B., Volpato, G. L., & Braithwaite, V. A. (2017). Physical and psychological motivation tests of individual preferences in rainbow trout. *Journal of Zoology, 302*(2), 108–118. https://doi.org/10.1111/jzo.12438

12. Morrissey, J. F., & Gruber, S. H. (1993). Home range of juvenile lemon sharks, *Negaprion brevirostris. Copeia, 1993*(2), 425. https://doi.org/10.2307/1447141

13. Finger, J. S., Dhellemmes, F., Guttridge, T. L., Kurvers, R. H. J. M., Gruber, S. H., & Krause, J. (2016). Rate of movement of juvenile lemon sharks in a novel open field, are we measuring activity or reaction to novelty? *Animal Behaviour, 116*, 75–82. https://doi.org/10.1016/j.anbehav.2016.03.032

14. Wolf, M., & Weissing, F. J. (2012). Animal personalities: Consequences for ecology and evolution. *Trends in Ecology & Evolution, 27*(8), 452–461. https://doi.org/10.1016/j.tree.2012.05.001

15. Linneweber, G. A., Andriatsilavo, M., Dutta, S. B., Bengochea, M., Hellbruegge, L., Liu, G., Ejsmont, R. K., Straw, A. D., Wernet, M., Hiesinger, P. R., & Hassan, B. A. (2020). A neurodevelopmental origin of behavioral individuality in the *Drosophila* visual system. *Science, 367*(6482), 1112–1119. https://doi.org/10.1126/science.aaw7182

16. Roland, G. (2020). The deep historical roots of modern culture: A comparative perspective. *Journal of Comparative Economics, 48*(3), 483–508. https://doi.org/10.1016/j.jce.2020.02.001

17. Ellickson, R. C., & Thorland, C. D. (1995). Ancient land law: Mesopotamia, Egypt, Israel. *Chicago-Kent Law Review, 71*, 321. https://scholarship.kentlaw.iit.edu/cklawreview/vol71/iss1/12

18. Malinar, A. (2015). Religious pluralism and processes of individualisation in Hinduism. *Religion, 45*(3), 386–408. https://doi.org/10.1080/0048721X.2015.1024038

19. Kateb, G. (2003). Democratic individualism and its critics. *Annual Review of Political Science, 6*(1), 275–305. https://doi.org/10.1146/annurev.polisci.6.121901.085719

20. Plato. (380 B.C.E.). *The republic: Book VIII.* Translated by Benjamin Jowett. https://www.gutenberg.org/files/1497/1497-h/1497-h.htm

21. Siedentop, L. (2014). *Inventing the individual: The origins of western liberalism.* Harvard University Press. https://www.hup.harvard.edu/books/9780674979888

22. Ehrlich, B., Silver, L. A., & Foot, J. (2024). History of Florence. In *Britannica.* https://www.britannica.com/place/Florence/History

23. The Editors of Encyclopaedia Britannica. (2024). Renaissance art. In *Britannica.* https://www.britannica.com/art/Renaissance-art

24. The Editors of Encyclopaedia Britannica. (2024). Leonardo Bruni. In *Britannica.* https://www.britannica.com/biography/Leonardo-Bruni

25. Schadee, H. (2018). A tale of two languages. Latin, the vernacular, and Leonardo Bruni's civic humanism. *Humanistica Lovaniensia, 67*(1), 11–46. https://www.jstor.org/stable/27172450

26. Skinner, Q., & van Gelderen, M. (2013). *Freedom and the construction of Europe* (M. van Gelderen, Ed.). Cambridge University Press. https://doi.org/10.1017/CBO9781139519281

27. Daub, S. (1996). *Leonardo Brunis Rede auf Nanni Strozzi: Einleitung, Edition und Kommentar* (Vol. 84). Walter de Gruyter GmbH. https://doi.org/10.1515/9783110954265

28. Hankins, J. (2006). Humanism in the vernacular: The case of Leonardo Bruni. I. In C. S. Celenza & K. Kenneth Gouwens (Eds.), *Humanism and creativity in the Renaissance: Essays in honor of Ronald G. Witt* (pp. 11–29). Brill. http://nrs.harvard.edu/urn-3:HUL.InstRepos:5479652

29. Bertram, C. (2003). *Routledge Philosophy GuideBook to Rousseau and the social contract.* Routledge. https://www.routledge.com/Routledge-Philosophy-GuideBook-to-Rousseau-and-the-Social-Contract/Bertram/p/book/9780415201995

30. Jones, C. (2017). *The smile revolution in eighteenth-century Paris.* Oxford University Press. https://global.oup.com/academic/product/the-smile-revolution-9780198715825

31. Kreitner, R. (2014, July 14). *Bastille Day, individualism and the concept of progress—in 1939.* The Nation. https://www.thenation.com/article/archive/bastille-day-individualism-and-concept-progress-1939/

CHAPTER 11

So Lonely: Isolation

I.

Being packed into Stonehaven with a bunch of other boats was not at all the escape that we had had in mind. So, for our final night of the charter, we returned to the one anchorage that we knew no one else would occupy. Because of the chaos of it, people avoid anchoring there. Funnel Bay.

We slowly motored into Funnel Bay through the narrow, deep channel in the middle. The water's surface was much calmer than the first time we anchored there. The bay was pristine. With no other boats in sight, we had our choice of prime anchorage. I found a spot on the chart that was as far into the bay as we could go. I wanted to leave some deep water on either side to allow for the boat to swing during the night—a lesson Funnel Bay had taught us on our first night.

We dropped anchor. This time we let out a lot of chain. Even after only 6 days of sailing, we were much more comfortable with the boat. It was a shame we would have to part with her. My husband stood at the bow and gave the hand signal to reverse so we could dig in the anchor. I kicked the gear into reverse, backing us up to a point where the boat no longer moved. My husband gave me a stop sign with his open palm and shouted, "Kill the engine." I pulled the fuel line, and the engine's rattling ceased. I was flooded with the stillness of the bay.

The sounds of chirping birds on the shoreline soon enveloped us. I took a deep breath, and I could actually hear it going in and out of my nostrils. *I just love being alone*, I thought.

The sun was beginning to dip behind the tallest trees on the western shore. The light hit the side of my face and nearly blinded me. It was the sun's last hoorah of the day, and the beauty queen she is, she demands that you pay attention. The only thing to do was to tolerate the intensity of the light and wait until the rays were eventually softened by the horizon.

We walked about the boat with our backs to the sun. All the features of the deck were distorted into blues and reds. We dutifully began to coil the ropes that lay sprawled around the cockpit. It was a solemn activity, coiling the ropes. It signified that our day of sailing was over, but at least we could still get some satisfaction in tidying up the last few affairs.

The water surrounding the boat began to take on hues of orange and pink. My brow finally relaxed as I settled into the cockpit. By then the sun was low enough that I could stare back at her, gazing across the glowing water. I sat frozen, mesmerized by it all.

I think if you're willing to go to places on Earth that no one else is willing to go to, your reward should be solitude. I have never climbed Mount Everest, but I can imagine that some of its first climbers reached the summit feeling the glory not only of conquering the mountain, but of being alone with the mountain. There is privilege in solitude. The problem is, solitude is bittersweet. No sooner do you get it, and it starts to feel...hollow. When I climbed Cradle Mountain (which, granted, is nothing like Mount Everest), I remember getting to the top, having that moment of solitude with the mountain, and then thinking, *Okay, now I want off this mountain.*

My eyes ran over the sleek lines of the boat. The neatly running ropes and polished stainless steel and carefully crafted bits of teak formed a work of art, possessing an orderly beauty akin to what I imagine Marie Kondo's closet must have—everything exactly where it should be. But I wondered, after Kondo's tidying is done and she has stared at her masterpiece for a while, is she not overcome with the meaninglessness of it all? Because even if you have everything in its place and you've managed to control all the chaos and tick all the boxes, you are still alone.

I descended to grab two beers from the galley fridge, and after my husband hung up the last neatly coiled rope, we cracked our beers to savor the last few drops of dusk. Aside from our "Cheers," we didn't say a word. I rested my feet against the center console. The white ropes that we had coiled on either side of the companionway had taken on the sunset's neon pink. I stared across the water. As the moments passed, its surface changed from magenta to periwinkle to deep grape. I looked up and saw the first stars appearing.

Darkness surrounded the boat. My beer tasted stale.

II.

It was December 2010 when I finally lost it. For the first few months, living in a sorority-type house in Miami was fun. We were only a few blocks away from the university's main campus. The air conditioner was always running, so you would be greeted with a refreshing wall of cool air every time you walked through the door. And our upstairs neighbors were bouncers, so they didn't mind at all when we came home late from clubbing, laughing at the top of our lungs.

The four of us got along for the most part. We would watch movies and cook together. Sometimes we would sit across from each other at the dining room table, with the Miami light pouring in over our cereal bowls as we poured our hearts out to each other about a recent heartbreak or a terrible grade on an exam.

The fellowship was good, but by the end of our first semester there, the cramped living conditions were starting to wear on me. The final straw was in December when everyone went home for Christmas. I was the only one who decided to stay at the house over the break, because my family was coming to visit me in Miami for Christmas. There was just one problem. One of my roommates had done a bunch of cooking before she left. I came home one day to pots and pans covered with pasta sauce and crumbs left on the stove and dirty dishes sitting on the counter next to the sink. Whichever roommate it was had left her dirty dishes to sit there in the kitchen for the entire month-long Christmas break. I snapped.

I was terrible at handling conflict. My first instinct was to pile all the dishes into a laundry basket and tape a note to it. I found a suitable basket, carried it to the kitchen, and pulled out a piece of paper and a pen. On the paper I drew little dishes and pots and pans crying, screaming

"Wash me!" I taped it to the front of the laundry basket, loaded all the dirty dishes in, and set it on the kitchen counter.

I remember staring at my "solution" through the kitchen doorway. As soon as my roommates saw this, I thought they would know that I was a terribly bitter and passive-aggressive person. Surely, I reasoned, the more adult thing to do in this situation would be to simply move out. So, following my family's visit, I packed up all of my belongings from that house and left. By the time that my three roommates returned from Christmas break, I was gone.

I felt emboldened by my decision. I was an independent woman, and I wouldn't be pushed around by anyone. My roommates were shocked by my decision. I assured them, in my typical avoidant style, that the move had nothing to do with them. I just wanted some space.

Space is what I got. I found a room for rent in a nice little cottage about a 10-minute drive from the university. The house was owned by a wealthy single woman in her 60s. It was immaculate and quiet, and in it I had to myself a wooden-floored bedroom with a big window and a bathroom that wouldn't be out of place in a luxury hotel. The woman's living room could've been Oprah's. The sofas were clean and white, with soft neutral throws draped over them. On the coffee table was a stack of books on food and art and travel that looked like they were hand curated by an interior designer. But the best part of all was that she was hardly ever there.

Living practically alone during that next semester, I experienced an incredible sense of quiet. I would lie on my bed for hours and just gulp up the sound of the breeze flowing through the palm trees outside my window. It was a well-shaded neighborhood, with hardly any traffic. Sometimes I would go for jogs in the morning and would wave at familiar walkers—healthy, affluent professionals in their 40s and 50s. It felt cordial, idyllic. It felt tidy. I wondered why I had waited so long to live on my own. For the first time in my life, I felt grown up. I loved it.

I was so taken with my new grown-up lifestyle that I hardly noticed the strange condition that was creeping over me. A few months into my new residence, I began to feel heavier. My body felt more sluggish to move around than usual. To combat this feeling of heaviness, I tried eating more vegetables and jogging more. But no matter how hard I tried to outrun it, the heaviness was always one step ahead of me. I began to feel tired all the time. It became harder and harder to get myself out of bed in the morning.

I would open my eyes each day, with a totally blank mind that was light as a feather, and then seconds later, a feeling of doom would descend on me like an elephant on my chest. It was as if I had suddenly remembered that someone close to me had died. I would begin to cry. I could never figure out why I was crying. No one had died. Everything in my life was fine, actually. And yet, with each passing day, life's inherent brilliance became fainter and fainter. *Is this what adulthood is supposed to feel like?* I wondered.

Before long, two vertical cracks formed between my eyebrows. They stayed there all the time, no matter how hard I tried to relax them. My face felt numb, like I no longer had the muscles to laugh or frown anymore. My skin became hardened like forgotten playdough. I wasn't ill, at least not by the conventional standards of illness, but I felt like I was deteriorating, day by day.

I couldn't for the life of me figure out what had caused my downturn. Here I was in this beautiful house with the total freedom to do what I wanted, and yet I was miserable.

Longitudinal studies have found that even more than age, gender, or the presence of life stressors, loneliness is the biggest risk factor for depression.[1] Loneliness warps the human mind, making it pay more attention to threatening stimuli and misperceive social encounters as threatening.[2] If left untreated, the effects of loneliness on the brain can cause people to withdraw even further, contributing to the development of negative thoughts and the onset of depression.[3] As early as 1962, depression was described as "a delusional idealisation regarding the worth of the self, marked by self-loathing."[4]

During that first semester of 2011, I started to feel disgusted with myself. I told myself that I was ungrateful, that I was lazy, that I was worthless. It felt like I was the only person in the world feeling this way. But in reality, I was hardly the only one feeling down. From 2010, happiness and life satisfaction took a sharp downturn among American adolescents. Prior to 2010, happiness had actually been on the rise in American adolescents since 1991.[5] But from 2010, it began to decline.[5] Meanwhile, rates of depression, suicidal ideation, and self-harm soared.[6] From 1999 to 2014, suicide rates in the United States tripled among girls aged 10–14 and increased among boys and older adolescents as well.[7]

This misery epidemic across the United States didn't quite make sense. The American economy was steadily improving from the Global Financial Crisis, income per capita was growing, and unemployment and crime rates were low. Compared with previous generations, binge drinking and heroin use were at historic lows among adolescents.[6] Yet something was definitely off with young people—their rate of depression was soaring. What was going on?

One possibility is that my generation was lonelier than previous generations. National surveys of 8.2 million American adolescents and students entering college conducted from 1976 to 2017 found that loneliness began to skyrocket around 2010.[8] Compared with previous generations, adolescents in the 2010s spent less time face-to-face with their peers, including getting together with friends, going to parties, going out, dating, going to the movies, and riding in cars for fun.[8] Students entering college in 2016 spent one hour less per day on in-person social interaction than did students entering in 1987.[8] The decline in social interaction was remarkable. Whereas in the 1970s over half of adolescents reported getting together with their friends almost every day, by 2017 only 28% of adolescents did the same.[8] This might explain why so many college students like me felt lonely.

The increasing loneliness of adolescents in the 2010s wasn't just an American phenomenon. According to the Organisation for Economic Co-operation and Development (OECD) Programme for International Student Assessment (PISA), the wellbeing of adolescents worldwide began to decline from around 2012. Between 2012 and 2018, school loneliness in 15- and 16-year-olds increased in 36 out of 37 countries studied. By 2018, nearly twice as many adolescents reported high levels of loneliness as in 2012. When analyzed, the loneliness was not significantly correlated with the countries' GDP, income inequality, or family size.[9] Something else must explain why, in the years following 2010, adolescents all around the world began to feel alone.

It seems like more than just a coincidence that the countries with the highest rates of school loneliness were also the countries with the greatest smartphone and internet access.[9] Smart phones became more prevalent in the 2010s, and by 2012 the majority of Americans owned one. Their use became so popular among adolescents that by 2017, the average 12th grader was spending over 6 hours a day on the internet, using social media, or texting. By 2018, 45% of American adolescents said that they were online “almost constantly.”[10]

At the same time that adolescents in the loneliest countries were on their phones and the internet, they were spending less time on activities like reading books and magazines and sleeping.[11, 12] Multiple studies have found that non-screen activities like face-to-face social interaction, attending religious services, and sleep are linked to greater happiness. On the other hand, studies that have followed the same individuals over time have found that digital media use predicts lower wellbeing later in life.[5]

It’s still not clear whether the rise in loneliness among American adolescents and students entering college around 2010 was caused by the increase in smartphone use or the decline in face-to-face interaction. Those adolescents with the highest levels of loneliness had the wicked combination of low in-person social interaction and high social media use.[8] It seems that starting in the 2010s, smartphones provided adolescents with the space, independence, and perception of control to be on our own and interact with others exactly as we pleased. But once we had this space, independence, and control, we began to suffer from the psychological effects of isolation.

I’m not trying to blame humanity’s loneliness on the acceleration of digital technologies or the advent of social media. I am pro-technology, and technology itself is never the root of the problem. The same Oldowan stone technologies that were used by our early ancestors to dismember animal carcasses so that food could be shared among tribes could have also been used to whack each other over the head. The problem is not the technology—it’s how we use it. Are we going to use social media to keep in touch with the people we’ve met and reach out to them years later, to stay up to date with our family, to share a laugh with our friends who are now scattered around the globe? Or are we going to use social media as a weapon, spreading fear and spouting our hatred of people with opposing philosophies? The choice of how we behave is ours, but it is not without consequences. Just as coming together has ensured our survival throughout human evolution, turning against each other can quickly lead to our demise.

III.

We know a lot about what isolation does to us emotionally from studies on sheep. Sheep are not the kind of animal to wear their hearts on their sleeves. But sheep have emotions too, and they hate being alone.

Staring a sheep in the face, you'd be hard pressed to discern how the sheep is feeling. For sheep, being emotionally expressive, like a dog, would call attention to themselves, and this is the opposite of what they want as vulnerable animals of prey. The best survival strategy if you're a sheep is to lie low and blend in with the flock.

Sheep are calmest when they are able to have physical contact with other members of the flock. If sheep band together in large groups, it becomes harder for a predator to take out any one individual. When sheep are grazing, they may spread out, but they always need to maintain visual contact with each other.[13] That's why when you separate a sheep from the sight of others, when you isolate them from the flock, they freak out.

Sheep are studied because of the reliability with which they freak out. To understand anxiety and what it does to the body, some researchers have used a measure of nervousness in sheep called the isolation box test. This involves getting animal ethics approval to place an individual sheep into a solid plywood box for one minute so that the sheep is unable to see any other sheep. In this situation, most sheep will become quite agitated. A digital agitation meter on the box keeps track of the sheep's movements and the number of high-pitched vocalizations (bleats) that the sheep makes while in the box.[14] The isolation box test poses no physical danger to the sheep, but most sheep, especially the ones with more nervous temperaments, go berserk in the box.[15] The isolation box test makes a reliable stressor for research purposes, because it consistently provokes anxiety in sheep. Individual sheep vary in their responses, but most sheep experience a panic response when they are isolated from other members of the flock. Such a panic response has an adaptive value for sheep—under natural conditions, it causes them to not veer too far from the flock, which stops them from being eaten.

Most animals hate being restrained, but for sheep in the isolation box test, something more is going on. Sheep are so emotionally dependent on each other, that even seeing a familiar face can ease some of their anxiety. Neuroscientist Ana P. da Costa and colleagues put individual sheep into an elongated isolation chamber for 30 minutes at a time. Upon entering the chamber, each sheep was shown four inverted white triangles for 15 minutes as a control. During these initial 15 minutes, the sheep took little notice of the triangles. Instead, they became anxious, with their heart rates increasing by about 20 beats per minute. The isolated sheep moved around in agitation and emitted several high-pitched bleats.[16]

After the sheep had been in the chamber for 15 minutes, the researchers changed the triangles to either the faces of three unfamiliar sheep or three goats. They then left the sheep in there for another 15 minutes. Meanwhile, the control sheep had to pass their time with just the same inverted white triangles. The researchers observed that those sheep who were given the faces of other sheep to look at spent a greater amount of time close to the pictures and looking directly at them, whereas the goat faces had no effect on the sheep. Furthermore, it would seem that seeing pictures of other sheep calmed the sheep, because their vocalizations and physical activity decreased significantly, and their heart rates returned to pre-isolation levels.[16] This is how we know that it's not just the restraint that sheep find to be anxiety provoking—it's actually the experience of being isolated from other members of their species.

The sheep in the isolation box test feel isolated even though they are not actually alone. They can hear other sheep through the box but cannot see them, so their *perception* is that they are alone. For these sheep, the mere perception of isolation is enough to provoke a stress response. It is the same with humans. Although we may ride the subway to work, spend all day helping customers, swing by a supermarket full of people on the way home, and spend time on social media all evening, we can still perceive that we are alone. Within the last ten years, researchers have started to ask, when an individual perceives that he or she is isolated, what happens to the body?

There is evidence that for social animals, isolation represents a potent stressor to the body, leading to increases in a key hormone involved in the stress response, cortisol.[17] In young pigs, a single isolation of just 4 hours was enough to produce a spike in cortisol concentrations.[18] In mice, social isolation increases their risk for obesity and type 2 diabetes.[19] Socially isolated mice are also more likely to have abnormal reactivity to stress, to develop cardiovascular disease, and to have exaggerated inflammation in their bodies.[20] These studies suggest that our hormonal response to isolation is not just negative in the short term but may lead to poor health in the long term.

It's not as easy to test the effects of social isolation on humans except in extraordinary circumstances, such as long-distance space flights, where humans are naturally isolated from each other. The National Aeronautics and Space Administration (NASA) conducted a study in which they isolated 16 participants for 30 days in the Human Exploration Research Analog, a closed module that simulates being in a spacecraft. Each participant was put in the module alone for 30 days, and at three points during the isolation period they were asked to pass their arm through a window to have their blood drawn. Another 17 participants were tested in the module but were not isolated. The NASA researchers found that the participants' cortisol levels increased significantly while they were isolated relative to their pre- and post-isolation blood levels, and their cortisol levels were significantly higher than the control participants who had not been isolated. A month of isolation was affecting the simulated astronauts' ability to cope with stress.[21]

The severe acute respiratory syndrome (SARS) pandemic in 2002–2004 was another circumstance of imposed isolation for humans. In Toronto, Canada, over 15,000 quarantined people were instructed not to leave their homes or have visitors. They were told to sleep in separate rooms, not share personal items, and wear masks when in the same room as other household members. After this, 129 people responded to a survey about their time in quarantine and following release. All survey respondents said that they felt a sense of isolation. Symptoms of depression were reported by 31.2% of respondents, and 28.9% experienced symptoms of PTSD. The prevalence of PTSD symptoms increased with longer durations of quarantine.[22]

We know that loneliness is a major predictor of depressive

symptoms,[23] but there is still a limited understanding of exactly how this happens. The neuroendocrine system, which is controlled by the brain, also acts on the brain itself, altering the way in which we perceive the world. Once an animal is depressed, the brain's perception becomes so warped that it becomes even more difficult to recover from an episode of perceived isolation. For example, 16 Merino sheep in Armidale, Australia, were given a drug to pharmacologically induce depression by decreasing their brain serotonin levels, while 16 control sheep were given an equivalent volume of saline. The sheep were then individually placed in a test arena with a threatening stimulus (an unfamiliar dog) outside the window. On another wall of the arena was a photo of another sheep, which the sheep could look to for comfort from the scary dog if they wished. But the researchers found that the chemically depressed sheep spent less time looking at the photo of the sheep and more time looking at the dog. They suggested that perhaps the depressed sheep had lost their normal coping mechanisms for stress or might have been experiencing social withdrawal, both of which are phenomena observed in depressed humans.[24]

If isolation results in the release of stress hormones, it is not surprising that social interaction provides some natural immunity to stress. Researchers have found that in mice, social contact increases the mobilization of the endogenous marijuana-like neurotransmitter, anandamide, in the part of the brain that regulates motivated behavior.[25] This neurotransmitter also signals for the release of the neuropeptide oxytocin, a hormone associated with social reward.[26] As we saw with the meerkats who sleep on top of each other, oxytocin is a key hormone that is released in response to physical contact with a trusted individual. It is critical for establishing maternal–infant attachments and other types of social bonds.[27] Oxytocin has also been shown to be involved in the development of trust, in pair bonding, and in the recognition of familiar individuals in rodents.[28]

Although physical contact was once thought to be necessary for the release of oxytocin, a study of 7 to 12-year-old girls and their mothers found that this hormone was released in the daughters when their mothers physically held them or merely spoke to them on the phone after the daughters had been through a stressful event. Remarkably, in both conditions, the girls' cortisol levels returned to normal whether through physical or voice contact, suggesting a stress-relieving effect of mother–daughter interaction.[28] However, when girls aged 7.5–12 were allowed to text message a parent after a stressful event, their parents' reassuring texts had no effect on their stress levels, similar to not having interacted with their parents at all. In the words of the research team, "We still need to hear each other."[29]

It turns out that social interaction is more narrowly defined by our mammalian bodies than it is by the smartphone-assisted standards of modern society. When we consider that the misery epidemic started at around the same time that smartphones became ubiquitous, it's easy to see that for social mammals like us, texting just doesn't equate to social interaction.

We may think that isolating ourselves from other members of our species causes no major harm. However, as we have seen from studies of other social animals, our brains would beg to differ. A growing culture of individualism has made us see ourselves as increasingly independent, each with our own desires and dreams to follow. Often it feels good to assert our independence and break away from social norms, living life on our own terms. Sometimes this requires moving far away from our family and friends. But isolation comes at a cost for social animals like us. We cannot easily escape the need for togetherness that has been written into our brains over millions of years.

IV.

For most of human history, the idea of intentionally isolating oneself would have seemed absurd. In the earliest human civilizations, community was the heart of everyday life. Our ancestors lived together, traveled together, slept together, and labored together. Although the ancient Greeks exalted the concept of the self, they believed that being part of a thriving community still took primacy over the self. When Socrates was condemned to death in Athens, he chose to accept his own execution rather than escape, as he believed that individuals who disobey the laws of their society tear away at the foundation of communal life. Even the pursuit of "knowing oneself" was not something that the ancient Greeks carried out in private; rather, it was a form of truth that was sought through dialogue and interaction with others.[30]

Advances in industrial manufacturing in Great Britain during the early 1700s injected new potential into humans' pockets and minds. Innovations like mechanized cotton spinning, steam power, the use of coke in iron production, and the first machine tools meant that products could be made with increasing speed and efficiency. These advances quickly spread throughout continental Europe and to the United States, starting the Industrial Revolution of 1760 to 1840. The increased output in production led to a surge in the GDP per capita in all regions of the world. Demand for labor in factories increased, attracting millions of rural residents to urban centers, where they could work for an hourly wage. In the process, individuals were uprooted from their communities of origin. All of a sudden, these individuals were forced to start life anew in faceless urban environments.

Most Americans lived in small towns until the late 1800s, working on small farms or in small businesses. In 1860 just one in five Americans lived in a city. But within just a few decades, more urban factories opened, and Americans left their farms in hoards to live and work in cities. By the start of the 20th century, Chicago, New York, Pittsburgh, and Milwaukee were growing, as millions of workers immigrated to work in industrial factories. Piled into jam-packed metropolises, these workers were disconnected from their communities of origin. By 1910, nearly half of the U.S. population were city dwellers. The pace of change was dizzying for American society. Journalist Walter Lippman observed in 1914, "We are unsettled to the very roots of our being ... We have changed our environment more quickly than we know how to change ourselves."[31]

New urban immigrants sought solidarity and redefined their communities by starting ethnic associations, professional organizations, hobby clubs, veterans' groups, lodges, and churches. Between 1870 and 1920, the number of new civic organizations in the United States skyrocketed. Although these clubs were generally exclusive to certain races and genders, new associations emerged from every corner of society.[31]

During the Great Depression, Americans were temporarily less likely to engage in their communities, but when the American economy improved during WWII and in the post-war economic boom, people established greater connections with their communities than they had in the preceding decades. According to American political scientist Robert

Putnam, "social cohesion," or the "connections among individuals—social networks and the norms of reciprocity and trustworthiness that arise from them," became stronger in the United States throughout most of the 20th century.[31] But by the late 1960s, the glue holding together the individuals in society had begun to lose its stickiness.

Putnam describes American community involvement over the 20th century as an inverted "U," rising from the early 1900s, pausing briefly during the Great Depression, and then rising to a peak in the early 1960s. From the 1960s, American engagement in civic associations and clubs declined. Church membership, which had previously been on the rise, began a sharp decline in the 1960s, plateauing in the 1980s and 1990s, before declining even further. Even at the family level, there was a decline in social cohesion beginning in the 1960s, with a waning percentage of middle-aged Americans being married.[31]

As traditional community structures fractured, so did social trust. During WWII, Americans were asked, "Generally speaking, would you say that most people can be trusted, or that you can't be too careful in dealing with people?"; 73% said that most people could be trusted. This percentage fell slightly in the post-war years before rising further to 77% in the late 50s and early 60s. From the mid-60s, however, the trend inverted, with social trust plummeting. By the early 80s, just under half of Americans said that most people could be trusted, and by 2015, only a third did.[31] Putnam has described the trend in American society over the last 125 years as a journey from "I" to "we" to "I."[31] Ever since the mid-1960s, Americans have been on a downward slope away from our previous sense of community, towards a society of isolated individuals. We seem to have fully bought in to the **Illusion of Separateness**.

In the second half of the 20th century, Americans were writing less about the groups to which they belonged and more about themselves as individuals. Sociologist Jean Twenge and colleagues analyzed over 700,000 American books published during 1960–2008 for the use of pronouns like I, you, me, we, and us. They found that between 1960 and 2008, the use of individualist pronouns (I, me) increased by 42%, while the use of collectivist pronouns (we, us) decreased by 10%.[32]

If you are like me, you're probably wondering, *What on Earth happened in the mid-1960s?*

When U.S. President John F. Kennedy (JFK) gave his speech at Rice University in September of 1962, he was clearly fueled by a cooperative hope for the nation:

> *We choose to go to the Moon in this decade and do the other things, not because they are easy, but because they are hard, because that goal will serve to organize and measure the best of our energies and skills, because that challenge is one that we are willing to accept, one we are unwilling to postpone, and one which we intend to win, and the others, too.*[33]

Just under a year later, Civil Rights Movement activist Martin Luther King, Jr., would deliver his "I Have a Dream" speech on the steps of the Lincoln Memorial.[34] Like JFK, he was filled with hope for a better future. And the quarter of a million people crowded in the National Mall to watch his speech believed that a better future was indeed possible. They believed in their communities. They believed in their nation. But in just a few short years, that belief would collapse.

It started just one month after JFK's optimistic Moon speech at Rice University. An American U-2 spy plane was taking a covert flight over the island of Cuba when they discovered and photographed nuclear missile sites being built by the Soviet Union. On October 16, 1962, President Kennedy was alerted by his National Security Advisor that a Soviet missile base was under construction just 90 miles from the coast of Florida and that the missiles had the capacity to reach major U.S. cities as far away as New York City and San Francisco.[35] After several days of intense meetings with his advisers, on October 22, JFK wrote a letter to the Premier of the Soviet Union, Nikita Khrushchev, reading "... I have not assumed that you or any other sane man would in this nuclear age, deliberately plunge the world into war which it is crystal clear no country could win and which could only result in catastrophic consequences to the whole world, including the aggressor." President Kennedy then informed the American public of his plans to establish a naval quarantine around the island until the Soviet Union agreed to dismantle the missiles and ship no further missiles to Cuba.[35]

With the quarantine established, President Kennedy was still concerned that a U.S. Navy vessel would be forced to fire upon a Soviet vessel as it headed into Cuba, which would have almost certainly sparked World War III. Therefore, on October 23 he sent another letter to Khrushchev, asking him to halt any Soviet ships headed toward Cuba. Khrushchev replied indignantly to President Kennedy's letter, accusing the United States of issuing an ultimatum and trying to intimidate the Soviet Union. On October 26, photographic evidence from American spy planes showed that the Soviets had accelerated the construction of the missile sites and were uncrating Soviet IL-28 bombers at Cuban airfields. On the same day, Cuban President Fidel Castro sent a private letter to Khrushchev urging the Soviet Union to initiate a nuclear first strike against the United States if the Americans invaded Cuba. The next day, an American U-2 plane was shot down over Cuba by a Soviet-supplied surface-to-air missile. The pilot, Major Rudolph Anderson, was killed. Many members of Kennedy's Executive Committee pressured him to take immediate military action against the missile sites in Cuba, but President Kennedy resisted. Instead, late on the night of October 28, President Kennedy's brother Robert Kennedy met secretly with Soviet Ambassador Anatoly Dobrynin, and they reached a basic understanding: The Soviet Union would withdraw the missiles from Cuba under United Nations supervision if the United States pledged not to invade Cuba and to eventually remove U.S. missiles from Turkey. The next morning, Radio Moscow announced that the Soviet Union had accepted the proposed solution, thereby putting an end to the 13 days that brought the world to the brink of nuclear annihilation.[35]

In the months and years following the Cuban Missile Crisis of 1962, Americans lived each day on edge. From the time she entered primary school, my mom in her little blond pigtails practiced "duck and cover" drills, where the students would rehearse hiding under their desks to protect themselves from flying glass and debris in the case of a nuclear bomb. With both the United States and the Soviet Union possessing enough nuclear power to completely destroy each other, any wrong diplomatic move could incite what would later be called "Mutually Assured Destruction."

If the threat of Mutually Assured Destruction by nuclear war wasn't enough to make Americans paranoid, the advent of television news reporting was enough to cement the paranoia. Black-and-white televisions were first introduced to the American public at the World's Fair in New York in 1939, but the TV sets were impossibly expensive. By 1963, however, 90% of American homes had televisions.[36] At 1:38 p.m. Central Standard Time on November 22, 1963, regular TV broadcasts in the United States were all interrupted by an urgent announcement.

Reporting directly from the TV newsroom where he had received the horrific news, CBS news anchor Walter Cronkite broke into tears on air. President Kennedy had been shot in his motorcade in downtown Dallas and was dead.[37] For the next 4 days, news coverage of the assassination rolled on Americans' living room television sets around the clock. Not a single commercial or entertainment program aired. At the time the White House confirmed Kennedy's death just after 1:30 p.m., 45.4% of American homes with a television had their sets in use. On Monday, soon after Kennedy's coffin arrived at Arlington National Cemetery, 81% of American homes with a television had their sets in use, one of the highest TV viewing rates in U.S. history.[36] According to Patty Rhule, a senior manager of exhibits at Newseum in Washington, D.C., "This is when America became a TV nation."[36]

Prior to JFK's assassination in 1963, print and radio were still the primary media for news reporting. But TV news coverage offered new opportunities for up-to-date coverage and on-the-scene interpretation of events that print and radio couldn't offer. As the 1960s progressed, Americans increasingly turned to TV news to find out what was going on in the world.[38] The death tolls of the Vietnam War, the Civil Rights movement and white backlash that sparked violent altercations in the American South, the assassinations of Martin Luther King, Jr., and of Bobby Kennedy, the Watergate Scandal—all of these events would be interpreted through the mouths of TV news reporters and witnessed through the glowing TV screen.

When our interpretation of the world is shaped through the selective lens of mass media rather than through face-to-face interactions with the people in our communities, the world seems much scarier than it really is. We could go so far as to say that the 1960s, with the advent of 24-hour TV news coverage, was the beginning of a "post-trust era."[39]

Seeing the terrible stories about the world coming through the TV, it's no wonder that we feel so alone. The mass media continuously reinforces our suspicion that humanity is headed towards Mutually Assured Destruction. Trapped inside this room of mirrors, it's so easy to forget that there is an alternative to Mutually Assured Destruction—one that we've been doing for millions of years...

CHAPTER 11 REFERENCES

Free Resources

Loneliness and social isolation—Tips for staying connected. (2021, January 14). NIH National Institute on Aging. https://www.nia.nih.gov/health/loneliness-and-social-isolation/loneliness-and-social-isolation-tips-staying-connected

Further Reading

Zelizer, B. (1992). *Covering the body: The Kennedy assassination, the media, and the shaping of collective memory.* The University of Chicago Press. https://press.uchicago.edu/ucp/books/book/chicago/C/bo3641671.html

References

1. Ypsilanti, A., Lazuras, L., Powell, P., & Overton, P. (2019). Self-disgust as a potential mechanism explaining the association between loneliness and depression. *Journal of Affective Disorders, 243,* 108–115. https://doi.org/10.1016/j.jad.2018.09.056

2. Cacioppo, S., Bangee, M., Balogh, S., Cardenas-Iniguez, C., Qualter, P., & Cacioppo, J. T. (2016). Loneliness and implicit attention to social threat: A high-performance electrical neuroimaging study. *Cognitive Neuroscience, 7*(1–4), 138–159. https://doi.org/10.1080/17588928.2015.1070136

3. Cacioppo, J. T., Hughes, M. E., Waite, L. J., Hawkley, L. C., & Thisted, R. A. (2006). Loneliness as a specific risk factor for depressive symptoms: Cross-sectional and longitudinal analyses. *Psychology and Aging, 21*(1), 140–151. https://doi.org/10.1037/0882-7974.21.1.140

4. Barry, M. J. (1962). Depression, shame, loneliness, and the psychiatrist's position. *American Journal of Psychotherapy, 16*(4), 580–590. https://doi.org/10.1176/appi.psychotherapy.1962.16.4.580

5. Twenge, J. M. (2019). The sad state of happiness in the United States and the role of digital media. In *World happiness report 2019.* https://sdghelpdesk.unescap.org/sites/default/files/2019-04/WHR19.pdf#page=89

6. Keyes, K. M., Gary, D., O'Malley, P. M., Hamilton, A., & Schulenberg, J. (2019). Recent increases in depressive symptoms among US adolescents: Trends from 1991 to 2018. *Social Psychiatry and Psychiatric Epidemiology, 54*(8), 987–996. https://doi.org/10.1007/s00127-019-01697-8

7. Curtin, S., Warner, M., & Hedegaard, H. (2016). Increases in suicide in the United States, 1999–2014. NCHS data brief, no 241. https://pubmed.ncbi.nlm.nih.gov/27111185/

8. Twenge, J. M., Spitzberg, B. H., & Campbell, W. K. (2019). Less in-person social interaction with peers among U.S. adolescents in the 21st century and links to loneliness. *Journal of Social and Personal Relationships, 36*(6), 1892–1913. https://doi.org/10.1177/0265407519836170

9. Twenge, J. M., Haidt, J., Blake, A. B., McAllister, C., Lemon, H., & le Roy, A. (2021). Worldwide increases in adolescent loneliness. *Journal of Adolescence, 93*(1), 257–269. https://doi.org/10.1016/j.adolescence.2021.06.006

10. Anderson, M., & Jiang, J. (2018). *Teens, social media & technology 2018.* http://assets.pewresearch.org/wp-content/uploads/sites/14/2018/05/31102617/PI_2018.05.31_TeensTech_FINAL.pdf

11. Twenge, J. M., Martin, G. N., & Spitzberg, B. H. (2019). Trends in U.S. adolescents' media use, 1976–2016: The rise of digital media, the decline of TV, and the (near) demise of print. *Psychology of Popular Media Culture, 8*(4), 329–345. https://doi.org/10.1037/ppm0000203

12. Twenge, J. M., Krizan, Z., & Hisler, G. (2017). Decreases in self-reported sleep duration among U.S. adolescents 2009–2015 and association with new media screen time. *Sleep Medicine, 39*, 47–53. https://doi.org/10.1016/j.sleep.2017.08.013

13. Schoenian, S. (2021). *Sheep behavior. Sheep 201: A Beginner's Guide to Raising Sheep.* https://www.sheep101.info/201/behavior.html

14. Murphy, P. M., Purvis, I. W., Lindsay, D., le Neindre, P., Orgeur, P., & Poindron, P. (1994). Measures of temperament are highly repeatable in Merino sheep and some are related to maternal behaviour. *20th Biennial Conference of the Australian Society of Animal Production, 20*, 247–250. https://research-repository.uwa.edu.au/en/publications/measures-of-temperament-are-highly-repeatable-in-merino-sheep-and

15. Hawken, P. A. R., Fiol, C., & Blache, D. (2012). Genetic differences in temperament determine whether lavender oil alleviates or exacerbates anxiety in sheep. *Physiology & Behavior, 105*(5), 1117–1123. https://doi.org/10.1016/j.physbeh.2011.12.005

16. da Costa, A. P., Leigh, A. E., Man, M.-S., & Kendrick, K. M. (2004). Face pictures reduce behavioural, autonomic, endocrine and neural indices of stress and fear in sheep. *Proceedings of the Royal Society of London. Series B: Biological Sciences, 271*(1552), 2077–2084. https://doi.org/10.1098/rspb.2004.2831

17. Cacioppo, J. T., Hawkley, L. C., Norman, G. J., & Berntson, G. G. (2011). Social isolation. *Annals of the New York Academy of Sciences, 1231*(1), 17–22. https://doi.org/10.1111/j.1749-6632.2011.06028.x

18. Tuchscherer, M., Kanitz, E., Puppe, B., & Tuchscherer, A. (2010). Altered immunomodulation by glucocorticoids in neonatal pigs exposed to a psychosocial stressor. *Pediatric Research, 68*(6), 473–478. https://doi.org/10.1203/PDR.0b013e3181f70f08

19. Nonogaki, K., Nozue, K., & Oka, Y. (2007). Social isolation affects the development of obesity and type 2 diabetes in mice. *Endocrinology, 148*(10), 4658–4666. https://doi.org/10.1210/en.2007-0296

20. Möller, M., du Preez, J. L., Viljoen, F. P., Berk, M., Emsley, R., & Harvey, B. H. (2013). Social isolation rearing induces mitochondrial, immunological, neurochemical and behavioural deficits in rats, and is reversed by clozapine or N-acetyl cysteine. *Brain, Behavior, and Immunity, 30*, 156–167. https://doi.org/10.1016/j.bbi.2012.12.011

21. Weber, J., Javelle, F., Klein, T., Foitschik, T., Crucian, B., Schneider, S., & Abeln, V. (2019). Neurophysiological, neuropsychological, and cognitive effects of 30 days of isolation. *Experimental Brain Research, 237*(6), 1563–1573. https://doi.org/10.1007/s00221-019-05531-0

22. Hawryluck, L., Gold, W. L., Robinson, S., Pogorski, S., Galea, S., & Styra, R. (2004). SARS control and psychological effects of quarantine, Toronto, Canada. *Emerging Infectious Diseases, 10*(7), 1206–1212. https://doi.org/10.3201/eid1007.030703

23. Hawkley, L. C., & Capitanio, J. P. (2015). Perceived social isolation, evolutionary fitness and health outcomes: A lifespan approach. *Philosophical Transactions of the Royal Society B: Biological Sciences, 370*(1669), 20140114. https://doi.org/10.1098/rstb.2014.0114

24. Monk, J. E., Belson, S., Colditz, I. G., & Lee, C. (2018). Attention bias test differentiates anxiety and depression in sheep. *Frontiers in Behavioral Neuroscience, 12.* https://doi.org/10.3389/fnbeh.2018.00246

25. Wei, D., Lee, D., Cox, C. D., Karsten, C. A., Peñagarikano, O., Geschwind, D. H., Gall, C. M., & Piomelli, D. (2015). Endocannabinoid signaling mediates oxytocin-driven social reward. *Proceedings of the National Academy of Sciences, 112*(45), 14084–14089. https://doi.org/10.1073/pnas.1509795112

26. Dölen, G., Darvishzadeh, A., Huang, K. W., & Malenka, R. C. (2013). Social reward requires coordinated activity of nucleus accumbens oxytocin and serotonin. *Nature, 501*(7466), 179–184. https://doi.org/10.1038/nature12518

27. Sussman, R. W. (2013). Why the legend of the killer ape never dies. In *War, peace, and human nature* (pp. 97–111). Oxford University Press. https://doi.org/10.1093/acprof:oso/9780199858996.003.0006

28. Seltzer, L. J., Ziegler, T. E., & Pollak, S. D. (2010). Social vocalizations can release oxytocin in humans. *Proceedings of the Royal Society B: Biological Sciences, 277*(1694), 2661–2666. https://doi.org/10.1098/rspb.2010.0567

29. Seltzer, L. J., Prososki, A. R., Ziegler, T. E., & Pollak, S. D. (2012). Instant messages vs. speech: Hormones and why we still need to hear each other. *Evolution and Human Behavior, 33*(1), 42–45. https://doi.org/10.1016/j.evolhumbehav.2011.05.004

30. *A brief exploration of history of western individualism.* (2010, April 5). Philosophy. https://philossophy.wordpress.com/2010/04/05/individualism/

31. Putnam, R. D. (2021). *The upswing: How America came together a century ago and how we can do it again.* Simon & Schuster . https://www.simonandschuster.com/books/The-Upswing/Robert-D-Putnam/9781982129156

32. Twenge, J. M., Campbell, W. K., & Gentile, B. (2013). Changes in pronoun use in American books and the rise of individualism, 1960-2008. *Journal of Cross-Cultural Psychology, 44*(3), 406–415. https://doi.org/10.1177/0022022112455100

33. Kennedy, J. F. (1962, September 1962). Rice University, 12 September 1962. https://www.jfklibrary.org/learn/about-jfk/historic-speeches/address-at-rice-university-on-the-nations-space-effort

34. King Jr., M. L. (1963, August 28). I Have a Dream. https://www.npr.org/2010/01/18/122701268/i-have-a-dream-speech-in-its-entirety

35. *The world on the brink: John F. Kennedy and the Cuban Missile Crisis, thirteen days in October 1962.* (n.d.). John F. Kennedy Presidential Library and Museum. Retrieved March 22, 2024, from https://microsites.jfklibrary.org/cmc/

36. Herskovitz, J. (2013, November 22). *How the JFK assassination transformed media coverage.* Reuters. https://www.reuters.com/article/us-usa-jfk-media-idUSBRE9AK11N20131121/

37. *November 22, 1963: Death of the President.* (n.d.). John F. Kennedy Presidential Library and Museum. Retrieved March 22, 2024, from https://www.jfklibrary.org/learn/about-jfk/jfk-in-history/november-22-1963-death-of-the-president

38. Zelizer, B. (1992). *Covering the body: The Kennedy assassination, the media, and the shaping of collective memory.* The University of Chicago Press. https://press.uchicago.edu/ucp/books/book/chicago/C/bo3641671.html

39. Public trust in government: 1958-2023. (2023, September 19). Pew Research Center. https://www.pewresearch.org/politics/2023/09/19/public-trust-in-government-1958-2023/

CHAPTER 12

What the World Needs Now: Cooperation

I.

Abell Point Marina was eerily still. Given the pearler of a sky and it being the start of the weekend, I expected to see people wheeling trollies down the marina arms, toting their canvas bags for the weekend and their ice chests full. But as we returned to the marina, no one was wheeling trollies. All the boats were boarded closed, the sail covers zipped tight. The boats were wiggling uncomfortably in their slips. The poor things were bound to stay like that until the pandemic was over.

We wheeled our trolley up to the marina office with our entire holiday's worth of stuff inside it. I carried the fishing rod in my left hand. It needed to be returned to the rental shop, along with our stinger suits, snorkels, and fins. My husband left me with the trolley in front of the rental shop and went to get the car.

Inside the rental shop, it was dark and cool. The man behind the counter asked me if we had a good time in the Whitsundays. "Yes, we did," I responded. The conversation was suspended there. Pushing the floppy stinger suits across the counter, I was reminded of returning my roller skates to the sweaty guy behind the skate counter at the end of every elementary school skate night. The lights would suddenly lift, the music would stop, and we'd roll to the benches to unlace our skates. The guy at the counter always insisted that laces be tucked inside. As you tucked the laces inside the boots and rolled them forward across the counter, you knew that the magic of the night was officially over.

After the snorkel gear, all that was left to return was the boat keys and access fobs. The charter office was on the second floor. I waited for my husband to pull up to the marina's cul-de-sac in his silver Subaru Outback. We transferred our bags from the trolley to the back of the car and then somberly climbed the stairs.

Everything about the marina seemed less shiny now that we were leaving it, even the stairs. The charter office spilled across the upper level to a full-length window, which overlooked dozens of masts against a blue sky. *It would be another perfect day for a sail,* I thought, *for someone who owns a boat.* Our status as "no-boats" never felt so poignant.

We approached the counter, where the young, bright-eyed, brunette receptionist looked genuinely happy to see us. Given how abandoned the marina was, I wondered if we were the first people to interact with her that day. She took a pump of hand sanitizer from a large bottle placed prominently on the reception counter. We followed suit. I proceeded to hand over the boat keys and the marina fobs.

The receptionist took an eternity to process our departure paperwork. The longer I stood at the counter with the boat keys gone from my hand, the more I felt like just an average schmuck. The only reason we could afford the charter is because the prices were so drastically slashed after people stopped traveling. You would think that after having had such a beautiful week island hopping around the Whitsundays, I would have the craving to do it again sometime. But in reality, I knew we may never do it again. It was a once-in-a-lifetime thing, and it was nice, but it felt unreal.

Behind us, a couple of men wearing polo shirts entered the charter office. They were tall and about our age. They had backpacks, although they looked too tidy to be backpackers. The receptionist was still engaged in whatever it was she was doing to process our checkout, so the guys took seats in the main lounge area of the office. They gazed out the window at the sky as if they were trying to read the weather.

My husband must have noticed them as well, because he turned to them and asked, "Headed out?"

"Ya," responded the taller of the two, "for the next three weeks." I ascertained either a German or a Dutch accent.

"Wow, three weeks. That will be fun," my husband responded. He was trying to sound genuine, but I could hear the envy in his voice. He looked back at me, and I raised my eyebrows.

The Dutch guy went on, "Ya, all the flights have been canceled, so there's nothing else to do really." He shrugged and chuckled to himself. *Nothing else to do.* I guess if you're a tourist with that much money, three weeks on a yacht is something to do.

Finally the receptionist released us. As we were leaving, I spotted a shelf next to the door bearing stubby holders with sailboats and the charter company's name on them. The receptionist followed my gaze, and for a second I almost thought that she was going to offer a couple of them to us as souvenirs, what with how much we spent on the charter. But when I inquired, she said they were $10 each. No free stubby holders for us.

We went downstairs, and my husband left to start the car while I wheeled the trolley back to the top of the marina ramp. As I parked the trolley, I made eye contact with one of the cleaners from the charter company, who was walking up the ramp, wheeling her cleaning supplies. She reminded me of my middle school band teacher—short and stout with spunky gray hair. She smiled at me with a cheeky, wide grin. My eyes gravitated towards her T-shirt. In big, black letters, it read "F*ck COVID." I laughed out loud. I was suddenly overcome with the urge to run down the ramp and hug her. I contained myself, instead just sending my smile of camaraderie. It felt good to know that, even amidst the fears of COVID, humor was still holding us together.

My mind flashed back to the motions of the two women fighting over toilet paper at the Woolworths in Sydney just a few weeks earlier. The viciousness in their eyes still haunted me. *Maybe there is a way to overcome all this fear*, I mused. If the woman cleaning the charter company's yachts was still coming to work each day with a smile on her face and donning a T-shirt message of resilience, maybe there is a way forward for all of us. Maybe it's in solidarity.

I returned to the Subaru, whose engine was already running. I opened the passenger door and stooped inside to sit on the cracked leather and wearing fabric seat. My husband had had the car for 7 years. It still smelled like toasted bread, the same as it did when he first picked me up from the airport in Brisbane.

I took a deep breath and shut my eyes. With all the travel lockdowns, I didn't know exactly how we were going to get back to Brisbane or what life was going to look like when we got there. But unlike our dream-like days on a luxury charter boat, that car felt real. It may have had a broken muffler, no central locking in two of the four doors, and worn seats, but unlike the boat we had been on for a week, our car was actually going to take us forward.

My husband connected his phone to the radio transmitter and pressed play before putting the car into gear. We pulled out of the marina driveway and turned onto the main road. As he accelerated, I heard the broken muffler roar, its familiarity oddly comforting. Through the speakers, the static at the start of the song transitioned into a melody. I smiled in recognition at my husband, and he grinned back at me.

"Mr. E's Beautiful Blues" by the Eels was the song that marked the start of our every adventure together. It meant we were about to roll up our pants and wade into the unknown. It meant that whatever came our way, we would face it head on, together.

Finally I came to know what the cleaner at Abell Point Marina already knew ... what perhaps we all know on a subconscious level. As humans, our life's purpose doesn't start when we run away from reality. It starts when we choose to face reality together.

II.

There was a jar on the bar at our house on Bird Road in Miami, labeled "Hopes and Dreams." It looked like something you would put on a little girl's dresser. As my friends and I sat around the table at Bird Road, doing our makeup before heading out to the clubs in Miami, we would often laugh about how out of place the jar looked there. Still, it must have made some impact on my psyche, because even now I think about that jar.

Today the only jar that I have is this book. It is a vessel for all my hopes and dreams for humanity. I think it's important that, even as adults, we protect our hopes and dreams, that we never let them die. Preserving our collective sense of hope is especially important when the outside world can seem so bleak.

The world seemed especially bleak at the start of 1989. The United States was already four decades deep into the Cold War with the Soviet Union. My mom recounted how she asked herself over and over, "How can I possibly bring a child into this world?" At that time, Americans were worried that a nuclear bomb could hit the United States at any time. I'm not sure how my parents made that decision to go ahead and have me in the midst of such an uncertain time. I don't know whether I would have been brave enough to do the same. But whenever I hear people say that today is the most uncertain time that humans have ever endured, I tell them what my mom told me. She took a risk by having me, despite no certainty about the future.

As we've seen throughout this book, chaos has been more the norm than the exception in the history of humanity. But we need not be paralyzed by it. Even during our most uncertain times, there are always tangible reminders of humanity's ability to cooperate.

The largest tangible evidence of human cooperation is directly above you. It orbits the Earth once every 90 minutes. The third brightest object in the sky, hovering 400 km above us, is a human-made structure the size of a football field, making it visible to the naked eye. It is comprised of metal from the United States, Russia, Japan, Canada, and several European countries. It's the International Space Station.[1,2,3,4]

When U.S. President Eisenhower signed the National Aeronautics and Space Act into law on July 29, 1958, it was decided that cooperation between the United States and other nations would be among NASA's primary objectives.[5]

NASA began designing a permanently staffed space station in 1984. Three separate orbital platforms were planned for research, observations, and deep-space exploration. At the same time, NASA reached out to international partners to foster cooperation. NASA signed agreements with the European Space Agency and Japan's National Space Development Agency, who provided research modules. In September 1988, the United States, Japan, Canada, and nine European countries signed an Inter-Governmental Agreement for the construction and utilization of the International Space Station.[6]

Meanwhile, the Soviet Union had been operating their own space station, *Mir*, and keeping crews in orbit for up to a year. But with the collapse of the Soviet Union in 1991, they were short on cash to continue their space projects. U.S. President Bill Clinton invited Russia to add the modules they had been planning for *Mir-2* to the International Space Station. Astonishingly, Russia agreed to the collaboration. In the years leading up to the International Space Station's launch, American astronauts were able to learn from Russian cosmonauts by joining them as long-duration residents aboard *Mir*.[6] Within just a few years, the two nations shifted their space efforts from Mutually Assured Destruction to Mutually Beneficial Cooperation.

The first modules of the International Space Station were launched in 1998.[6] To date, this multinational station is the largest single structure humans have ever put into space.[1] There are 15 nations that contribute to the International Space Station, with no single nation owning it. As of March 2024, 280 individuals from 23 countries had visited the International Space Station.[7]

Although political tension remains between Russia and the United States, heightened most recently when Russia invaded Ukraine, it may be helpful to remember that cooperation among feuding nations can sometimes be initiated by a single individual taking the first step towards cooperation.

In 1977, Egyptian President Anwar Sadat became the first Arab leader to visit Israel. Egypt and Israel had been in conflict for the 30 years since Israel's foundation, and just 4 years earlier, Sadar had sent Egyptian troops into the Sinai Peninsula, placing the two countries at war.[8] U.S. President Jimmy Carter was convinced that there would be no movement on peace in the Middle East without the United States orchestrating it.[9] Carter began arrangements for an international peace conference in Geneva, which would bring together the United States, the Soviet Union, Israel, the Arab states, and the Palestinians, with the intention of solving the Arab–Israel conflict once and for all.[9]

But on November 9, 1977, Egyptian President Sadat shocked the international community when he addressed the People's Assembly in Cairo, announcing he would travel to Jerusalem to expedite peace talks.[9] Unbeknownst to the United States, secret discussions had been going on between Israel and Egypt for some months. In August the Israeli Prime Minister Menachem Begin had flown to Bucharest to ask Romanian leader, Nicolae Ceauşescu, to convey a message to Sadat that Israel was serious about a peace accord. Following this, Begin's foreign minister flew to Morocco for secret talks with Egypt's deputy prime minister. During these talks it was understood that Jerusalem would be flexible on the most important question of territory.[9] Two days after Sadat's public announcement of his willingness to travel to Jerusalem, Begin addressed the Egyptian people, welcoming Sadat's visit to Jerusalem and pledging that he, for his part, would be ready to come to Cairo. He said, "No more wars—peace—a real peace, and forever."[10]

As Sadat's motorcade moved towards the Knesset in Jerusalem, thousands of ecstatic Israelis lined the streets to welcome a man who, just 4 years earlier, was their sworn enemy. For the duration of Sadat's trip, Israelis were in awe that the leader of the most powerful and populous Arab state had come to make peace.[9] On November 20, Sadat and Begin held a joint press conference in Jerusalem, in which they announced an end to hostilities between their nations and their pledge to work towards peace.[8] This announcement formed the first stepping stone to the Camp David Accords in September 1978 and the signing of the Egypt–Israel Peace Treaty in March 1979.[8] In October 1978, Sadat and Begin were awarded the Nobel Peace Prize—just 3 years before Sadat was assassinated by extremists opposed to the peace treaty.[8] This story demonstrates just how difficult it can be to achieve lasting peace in some regions but also how powerful it can be when our world leaders set the example by choosing to cooperate first.

The use of cooperation as a way of solving problems has been applied not just at the political level to resolve conflicts among nations but also within the capitalist arena of business.

In 2001, Yvon Chouinard, founder of Patagonia, and Craig Mathews, founder of Blue Ribbon Flies, were fishing along the banks of the Madison River when they came up with an idea: *What if we committed to giving 1% of our companies' sales back to the planet, regardless of whether or not we are profitable?*[11]

And so, 1% for the Planet began, a global movement inspiring businesses to give 1% of their gross sales each year to supporting environmental solutions. Soon many other companies besides Patagonia and Blue Ribbon Flies realized that because they profit from the resources they take from the Earth, they have a responsibility to protect those resources. Brands like Brushfire Records, Klean Kanteen, New Belgium Brewing, and Caudalie joined the movement, followed by over 5,000 businesses across 65 industries worldwide.[11, 12, 13] Co-founder Yvon Chouinard says of the movement, "This is not philanthropy. This should be a cost of doing business. It's paying rent for our use of the planet."[14]

To date, 1% for the Planet partners have given over $600 million to the environment,[11] channeling it into causes like creating the first regenerative cotton farm in Turkey (Organic Basics & WWF), preserving 14,000 m^2 of land in Germany as a wild bee playground (Green Forest Fund & 3 Bears),[15] and investing in women's clean energy businesses in off-grid communities across Africa (Solar Sister).[16] Although the partners in 1% for the Planet may be competitors in business, they have chosen to join efforts in pursuit of the noble goal of saving the planet.

Reading these examples, it would be easy to assume that you need a lot of power or money to make a difference in the world using cooperation. But my favorite example of cooperation shows otherwise.

When ocean explorer Jacques-Yves Cousteau visited the Sea of Cortez in the late 1980s, he described it as "the world's aquarium."[17] This 1,200 km long stretch of ocean cradled between the peninsula of Baja California Sur and the west coast of mainland Mexico is home to almost 900 species of fish and roughly one third of the world's marine mammals, including 11 species of whale.[18] It wasn't always this way, though. Beginning in the 1960s, when commercial fishing began in the region, fish stocks and populations of keystone species like sharks began to plummet.[19]

At the time, the Mexican government's regulations to prevent overfishing were not being enforced in the Sea of Cortez. With the Sea of

Cortez as a free-for-all for commercial fishing industries, the small region became the source of half of Mexico's fishing economy.[20] Consequently, local fishers soon struggled to find fish that were once abundant in the Sea of Cortez.

Seeing their marine environment destroyed, and with the government doing little to protect the waters, locals of the small coastal village of Cabo Pulmo took matters into their own hands. In the early 1990s, at the encouragement of local fisherman and scuba diver Mario Castro, around 100 residents of Cabo Pulmo who relied on fishing for their livelihood put their heads together to come up with a solution.[21]

The community of Cabo Pulmo decided that they would stop fishing the area. Although fishing was their livelihood, they all agreed to let the area lie fallow for several years so that the fish stocks and marine life could recover. In an interview with science writer Tim Folger, resident Paco Castro Lucero explained, "For the first five years it was very hard for our families, because we used to do commercial fishing." But the locals were convinced that if the marine life were allowed to recover, the region could one day become a booming destination for ecotourism. Snorkeling and scuba diving might bring as much, if not more revenue, than fishing once had.[20]

At the same time, the community began petitioning the federal government to establish a marine reserve at Cabo Pulmo to protect the region from commercial fishing. In June of 1995, the Mexican government granted the community's request, establishing the Cabo Pulmo National Park.[22] Gill nets, trawling, and long lines were banned throughout the 71.11 km^2 park.[18, 20] In addition, 35% of the park was designated as a no-take zone, where all fishing was prohibited.[22] Many of the families in Cabo Pulmo believed that the no-take zone was too small, so they decided on their own to adhere to a no-take policy throughout the whole park. They began patrolling the waters of the park in their own boats to guard the area against poaching.[20]

Within a decade, the cooperation of the community at Cabo Pulmo was already paying off. Coral reefs throughout the Cabo Pulmo National Park recovered. Fish biomass increased by 460%, restoring the area to a biomass similar to coral reefs that have never been touched by fishing. Populations of whale sharks, manta rays, humpback whales, and endangered sea turtles rebounded.[22] Marine conservationist Enric Sala said, "[Fish] were so abundant that we could not see each other if we were fifteen meters apart. We saw more sharks in one dive at Cabo Pulmo than in 10 years of diving throughout the Gulf of California!"[22]

As the community had hoped, Cabo Pulmo's economy also rebounded. In an interview with WWF International, Judith Castro said, "Now many of the women in Cabo Pulmo are business owners. They have opened their own businesses just like their husbands. The fishermen now have their diving or snorkeling shops, and the women have started their hospitality businesses, sale of artisanal products, souvenirs, and restaurants ... It fills us with optimism. It fills us with pride."[23] Traditional ways of fishing have also recovered. Cabo Pulmo's fishermen can now make a living by fishing outside the park's boundaries because the park is so abundant that the fish spill out into other areas.[23]

These stories of cooperation are not just meant to be quaint tales to make us feel better about the world. They carry a much stronger message. As Judith Castro of Cabo Pulmo put it, "The message is that it can be replicated, that there can be lots of Cabo Pulmos all over the world. And not only that there can be, there *should* be, there *needs* to be."[23]

III.

So, if cooperating through chaos is possible for humanity (and it clearly is), then how do we begin?

As always, I turn to animals for direction. Elephants (*Elephas maximus, Loxodonta africana, and L. cyclotis*) have some of the largest social networks of any non-human animal. And within these social networks, elephants are good at cooperating for many aspects of their survival, including group defense, resource acquisition, and care of offspring. Their high degree of cooperation requires elephants to build strong and lasting relationships.[24] As with humans, the strength of our relationships is often tested the most when we're facing the unknown. While at Melbourne Zoo in November 2019, I was able to see how groups of elephants deal with the unknown.

I was there to meet with the zoo's Director of Wildlife Conservation and Science, Sally Sherwen. Sally has a background in animal welfare research, and since beginning work at Zoos Victoria, she has done some interesting things to bring scientific research on animal welfare into practice in the zoo setting. That day, Sally was on her way to the elephant enclosure, so she offered to introduce me and my colleague to the Elephants Life Sciences Manager, Erin Gardiner. She walked us past the orangutan enclosure and deep into what appeared to be a tropical Asian village. Before any sign could give it away, I noticed the distinct smell of elephants wafting through the palms.

As we approached, a zookeeper with her brunette hair in a ponytail was emerging from the back of the elephant enclosure. Sally introduced us to Erin, who smiled as she shook our hands. In her other hand she held an orange plastic bucket. It swung to and fro as she led us to the front of the fenced enclosure, where there was a large swimming pool and a spraying fountain for the elephants. Sally left us there with Erin, who proceeded to explain the zoo's elephant enrichment program.

The elephants must have recognized Erin, because eventually all six of them marched over to congregate around her. The elephants approaching us were Asian elephants, with rounded heads and droopy ears. To maintain their cooperative relationships, elephants must have an amazing social memory. Ethologist Karen McComb and colleagues studied elephants' recognition of voices that were played back to them and concluded that female elephants can remember the voices of least 100 other adult females.[25] So I had no doubt that these elephants recognized Erin.

As Erin spoke, she pulled some greens out of the orange bucket and passed them to the elephants' grasping trunks. "So, you'll see the exhibit looks a bit messy at the moment," she said. "We are in the process of adding five new tree trunks ... We've had construction crews and cranes coming in to drop them. The idea is that they'll provide scratching opportunities for the elephants and that they'll spend time trying to get the bark off, which will grind down their tusks. We're also in the process of installing some tall poles so that we can hang up large toys for the elephants to interact with."

I looked up at the poles, which towered above the elephants' heads. Next to one of the elephants, I noticed the largest rubber ball I had ever seen. "Where do you even get elephant toys?" I asked in disbelief. Erin chuckled. "There is one company in Australia that makes them, but most of our toys have to be custom made. And a lot of our enrichment uses natural materials, like tree trunks or different herbs that we scatter around. The zoo can become a static environment if we don't provide a lot of different enrichment opportunities, so we really have to get creative." She grinned.

The elephants I've seen have always seemed quite curious, especially the young ones. When elephant calves are just 1–2 months old, they begin to sample potential food items by grasping objects with their trunk, rolling things around, and even placing their trunks in the mouths of adults and pulling out food items and eating the fresh dung of other elephants.[24] I guess if you're a young animal who can eat such a wide variety of food sources, it helps to learn directly from the experiences of others.

Erin continued, "To know that the elephants here are having the same opportunities that they would have in the wild, we want to see them active, we want to see them investigating their environment, and we want to see them working as a cohesive herd."

In wild settings, even the diversity of social interactions that elephants have likely forms part of their daily enrichment. Similar to us, elephants live in fluid, multi-tiered, fission–fusion societies where group membership changes frequently. It is thought that wild elephants living in large populations likely encounter hundreds of other individuals each day.[24] Still, elephants are one of the only animals on Earth (apart from lions, hyenas, sperm whales, chimpanzees, and humans) known to form such close and lasting relationships, despite having fluid social systems.[26]

Erin said, "One of the challenges of a zoo environment is the predictability, which is something that the elephants wouldn't have in the wild. Elephant herds in the wild will come across a lot of new things, whether it is a crocodile in the river or a patch of particularly rich food, so we need some way of simulating this lack of predictability here in the zoo."

"And how do you do that, in a practical sense?" I asked.

She answered enthusiastically, "One of the things that we've done recently is to bring different musicians in to play for the elephants. Elephants have really good hearing," she said, "and we wanted to see how they would respond to live music. We wanted to see their reaction to different sounds. And what we really wanted to see is whether these unpredictable experiences could elicit natural behaviors in the herd."

My eyes widened. "What did they do when they heard the music?"

Erin giggled. "It was actually amazing. There had been some other studies of people playing live music to animals in the zoo, but we had no idea what to expect for elephants. They reacted to the different styles of music really differently. During the classical cello concerto, they were quite gentle, and two of the elephants actually intertwined their trunks. And then some of the percussion instruments elicited these really beautiful herding behaviors. They all moved close together and touched each other, flapped their ears, and made vocalizations. Even after the music had stopped, they remained cohesive for the rest of the day."[27]

I was fascinated. "Why do you think that was?" I asked her.

"We consider this to be a positive social behavior. They're a really strongly bonded unit, a matriarchal society, so they were responding to the music as a herd. Elephants use infrasound, so they could have been responding to the vibrations of the drums, but what we think they were doing is they were viewing that as an unknown stimulus in their environment. They had never heard anything like it before, so they were trying to assess whether it was positive, negative, or neutral and communicate to each other about how they were going to respond."[28]

I found it amusing that the keepers at Melbourne Zoo needed to artificially create a bit of chaos for the elephants. But I suppose if these animals normally use their herding behaviors to respond to the unknown, they needed some sort of unknown to trigger those cooperative behaviors. It sounded as if the foreignness of the percussion music did the trick. The unpredictable element in their environment—that dose of chaos—had made them work together.

Comparative psychologist Joshua Plotnik and colleagues have found that elephants are so attuned to working with each other to solve problems that they actually know when they are going to need help to solve a problem before they begin to do so. When elephants were given a cooperative task requiring two individuals to simultaneously pull two ends of the same rope to obtain a reward, the elephants not only worked together on the task—they actually waited for up to 45 seconds before pulling if their partner wasn't there yet.[29] They seemed to have understood that without the help of a partner, this task was going to be too much for them to carry out on their own.

There is also evidence that elephants respond to stressful situations by comforting each other. In a 2014 study carried out at an elephant camp in Thailand, Plotnik and ethologist Frans de Waal found that when an elephant experienced unseen or negative stimuli and displayed distress behavior (vocalizations and body postures signaling agitation), the distressed elephant was often approached by a bystander elephant who initiated reassuring contact, including vocal communication and trunk touches to the genitals, mouth, and the rest of the head. The strong response of the elephants to a herd member in distress indicates that elephants pay attention to each other's emotions and respond to each other with consolation.[30] Just as Erin had described at Melbourne Zoo, in the Thai elephant study, after 30 of the 42 distressing events, bystander elephants formed a close circle around the distressed individuals—an anti-predator behavior known as bunching.[30]

At this stage, Erin's orange bucket had run out of food, so the elephants dissipated. I stood in silence as I watched their boulder-like bodies shuffle away.

When Erin mentioned that the elephants were communicating about how they were going to respond, she was referring to a form of cooperative decision-making that is thought to occur among elephants. Based on their opportunistic observations of the vocal exchanges that occur between elephants, J. Poole and C. Moss reported that elephants make decisions about things like how to respond to a predator or where they are going to go using participatory decision-making sessions, which involve consensus building through vocal exchanges that may last up to an hour.[24]

Of course, unless you are standing very close to a herd, you will not hear elephants' intimate discussions. Most of their vocal communication takes place below the 20 hertz low-frequency limit of human hearing, in the infrasonic range. In addition to the well-known trumpets and screams that you might have heard elephants make at a zoo, elephants have a large vocal repertoire of infrasonic "rumbles" that they use to communicate with each other.[31] One advantage of communicating at very low frequencies of sound is that the vocalizations can be heard over long distances, a property of sound that whales also use to communicate across the ocean.

Could it be that elephants handle uncertainty in their environment by coming together to discuss their options? If so, perhaps we should take a lesson from the elephants and come together to discuss our options.

IV.

When it comes to the many challenges facing our species now, they are burdens far too large for any one of us to carry on our own. There is a solution to these challenges, and it doesn't involve destroying each other. Throughout this book we have seen how cooperative behavior has helped our species, like many other animal species, to survive times of great environmental tumult. We have reviewed the compelling evidence that humans evolved to be a highly cooperative species. Our cooperative instincts are written into our DNA. Cooperation has always been, and continues to be, the glue that binds humanity.

Humans aren't only under the influence of our genes, however. Cultural evolution—which includes our knowledge, beliefs, and the ideas that we subscribe to—has the capacity to change the trajectory of our species, and in many ways, it has. In this book I have proposed three illusions that I believe might have changed the trajectory of our species, moving us away from our cooperative instincts. Illusions are merely cognitive constructs, so others may define them differently, but I think they are helpful constructs for the purpose of examining where humanity needs to head next.

The **Illusion of Security**, which could have emerged when our ancestors began to stockpile resources to protect themselves against times of scarcity, might have made humans less reliant on each other for our survival. Prior to this illusion, our ancestors needed to cooperate in order to acquire food, rear their young, and protect themselves against predators. Stockpiling, storage structures, and walls to protect surplus resources and themselves would have made the need to cooperate seem less pressing, giving our ancestors the false belief that they were safe.

Of course, natural disasters like floods and bushfires, as well as human-made disasters like war, prove that we are never truly secure, no matter how much money we have or how strong of walls we build.

The **Illusion of Superiority**, which could have emerged when certain humans had differential access to resources or were given positions of prominence within their communities, might have led us to believe that some people are better than others and deserve more than others. This illusion, which might have been connected with early forms of religion or the first semi-permanent settlements, could have been the start of some of the massive inequalities that we see in the human population today. Under this illusion, it is difficult for humans to share resources, trust each other, and cooperate. Of course, it only takes something like a global pandemic to remind us that we are all susceptible to the forces of nature and are constrained by our common human biology.

The **Illusion of Separateness**, which could have emerged when humans first took up arms and went to war against their neighbors, might have heightened our fear and distrust of people who do not look or talk like us. Perhaps this illusion led to the warring society that we have today, where we look at our enemy as if they were a fundamentally different or non-human entity. Of course, we are now beginning to realize that even as separate tribes and nations, we have all contributed to the destruction of our common habitat, planet Earth, and we are all suffering the effects of this destruction, as are other animals who had nothing to do with making the mess.

These three illusions are the products of cultural evolution, not the products of genetic selection, so the good news is that they can be changed fairly easily by further cultural evolution. Cultural evolution happens quickly compared with natural selection—our entire way of thinking can shift within a few decades. In fact, we have witnessed this transformation happen many times. My mother's generation experienced a wave of female empowerment. Whereas women in her mother's generation mostly stayed home to raise children, many women in my mother's generation completed university degrees, became working professionals, entered politics, and took leadership positions. As I was coming into adulthood, there were already several female presidents and prime ministers of nations around the world, including here in Australia. Our cultural belief about what women are capable of has shifted within the course of just three generations. Similarly, I have seen the cultural evolution in our society's beliefs about animals change dramatically over the course of my own lifetime. When I was a child, almost no one believed that animals were sentient, or feeling, beings. Today most scientists and a large part of the public believe that animals experience suffering, fear, pleasure, and joy. Our societal awareness about animals and our concern for their welfare continues to shift along a positive trajectory. If these cultural illusions that women couldn't become professionals and that animals couldn't feel have changed in less than a century, then why can't the illusions of Security, Superiority, and Separateness shift too?

Cultural anthropologist Margaret Mead once said, "Never doubt that a small group of thoughtful, committed citizens can change the world. Indeed, it's the only thing that ever has." We don't need all of humanity to change their beliefs at once in order to move the needle in the right direction. In fact, as Margaret Mead said, it only takes a few of us to start a change.

I encourage you to ask yourself what change you wish to see in the world and to put that wish into cooperative practice. We each feel passionate about certain causes over others because of our unique personalities and life experiences. We are all pulled towards something. I believe that if you listen to that pull and you find others who are being pulled in that same direction, you can cooperate to create a positive change in the world. There is no doubt in my mind that every human alive has the power to do this.

When I was first working out how to cooperate with others to shape a better world for animals, it would have been so helpful to have a practical guide in front of me, showing me how to approach it. Hence, the Seven Steps to Cooperation that follow—hopefully they can be helpful to you. These practical steps may read like a recipe, but I'm not suggesting that you follow them to the T. They are merely suggestions to get you into the kitchen, pulling out the pots and pans. After you've started, and there's pasta sauce splattering on the walls, please adapt this recipe to your own taste.

I cannot promise you that by following the Seven Steps to Cooperation you will find the answers to humanity's greatest challenges. I can only say that these are the steps that I would have liked to have had in front of me when I first started.

1. FIGHT THE BATTLE WITHIN

Humans are creative beasts. Like many other animals, it is in our nature to explore, to experiment, and to experience. But humans and other animals are also loath to exert energy, especially if the payoff is distant or uncertain. I refer to these two animal impulses as creativity and resistance, respectively. Others may refer to them as approach and withdrawal, opening and closing (think sea anemone), action and inaction, energy and friction, or risk and safety. Like a seagull who gets fed chips in the same spot every day, we would much rather stick with what is comfortable and predictable than use our creative brains to adapt to something new. Therefore, the journey to do anything positive out there in the world begins within our own brains.

We all have creative ideas every day, but what stops us from pursuing them is resistance. It is an internal force that holds us back, preventing us from ever taking the first step. Creativity and resistance are as much a part of the animal world as acceleration and inertia are part of the physical world. There is nothing inherently good or bad about either of them. They work together like the accelerator and the brake on a car. Without one, the car never moves. Without the other, the car drives off a cliff.

For me, resistance shows up like clockwork. Any time I attempt something new or difficult, whether it be writing, painting, scuba diving, sailing, traveling, building a website, delivering a talk, meeting with a new client, preparing a report for the first time, or making a phone call, resistance wells up inside me, trying to stop me in my tracks. Reading the life-changing book *The War of Art* by Steven Pressfield has taught me that this resistance will never go away—the only way to push past it is to take one small step in the right direction. And then to take another small step. And then another. In the philosophy of the book *Bird by Bird* by Anne Lamott, you can overcome resistance daily by concentrating very narrowly on the next step right in front of you and, almost without thinking, doing it. Having done this for the past couple of years, my life has changed dramatically.

I believe that if we could all make overcoming resistance a daily practice, as a species we could unleash more creativity and potential for positive change than humanity has ever seen.

2. SHAPE YOUR VISION

Animals cooperate for a common purpose, whether it be vigilance against predators, raising their young, gaining access to a food source, or deciding how to respond to chaos. Of course, animals do plenty of socializing as well, just hanging out in groups, but cooperation is different to just hanging out. Cooperation is when animals work together towards a common goal. It requires joint intentionality—all the animals have the same vision of what they want to achieve. I have seen many organizations where managers bemoan that employees will not cooperate. But if there is no shared vision, no common purpose, then why should animals cooperate?

The shared vision can be anything. It can be anything that you really care about and that you suspect other people might care about as well. Improving the welfare of animals, giving children a safe and stress-free environment to grow up in, saving the ocean's habitats from degradation, reversing climate change, giving people mental health support networks, enabling the elderly to pass on their wisdom, bringing together professionals to share their knowledge ... the possibilities for a shared vision are endless. The main thing is that the visionaries care deeply about it. Because when the road to building cooperation gets difficult (and it definitely will), your commitment to your vision is the only thing that's going to keep you going.

As we've seen with the Nazi and white supremacy movements, the shared vision doesn't necessarily have to be a morally good one to get people gathered around it. But I would encourage you to pursue a vision that all of humanity can get behind, not just certain races or classes. Sowing the seeds of hatred and division is a recipe for further societal fragmentation, which in the end just hurts all of us. What humanity needs now is not more tearing apart but coming together.

Modest visions tend to get modest support. It's simply not worth people investing their effort for only a moderate gain. The more ambitious the vision, the more people are drawn to it. I like to pick up things that have been put in the "too hard" basket. They make excellent candidate problems for cooperation.

To achieve cooperation, you need a goal that is so ambitious so as to be inspirational—hearing it spoken should give you goosebumps. The reason that you need this ambition is because humans, like other cooperative species, are wired to pull together in the face of adversity. If there is no adversity, no challenge, then there's no reason to cooperate.

I've found that if you can articulate your vision in three to six simple words, you'll have a better chance of success. To me the best example of a vision was that articulated in JFK's moon speech: "Put a man on the moon." That is a simple and concrete vision. Anyone can understand and picture it. Yours could be to "Put a roof over every head in Chicago" or "Clear the ocean of plastic." The clearer your vision, the better your chances will be of people working together towards it.

3. SHARE YOUR VISION WITH PEOPLE

As social animals, we learn almost every basic skill that we need for our survival through social learning. We observe others and imitate their behavior, or they instruct us directly. Why, then, when I was in college, did I try to start a non-profit organization to sell fair trade clothing without first talking about the idea with other people selling fair trade products? Maybe it was adolescent arrogance, or maybe I was afraid that someone was going to steal our idea. Trust me, building a cooperative movement requires a lot of work. No one is going to steal your idea. As Benjamin Franklin said, it's 1% inspiration and 99% perspiration. Only you are passionate enough to put in the work required to make your vision a reality. You have nothing to lose and everything to gain by talking with other people who are already doing something similar.

Why do you need to find people who are already doing something similar? Because they already share your vision. You may think that you could start by sharing your idea on social media or talking about it with your friends and family. They may wish you the best, but your friends and family may not know whether your vision is worth pursuing, whether you could join someone who is already doing it, or whether similar attempts have failed in the past. The only people who know this are those who are in the arena, doing this work, day in, day out.

So, if your vision is "Make the ocean safe for whales," go talk with the Sea World Foundation, talk with Sea Shepherd, talk with the World Wildlife Fund, talk with the Ocean Alliance. Talk with the people who have been doing this for decades. They will have the wisdom that you need to get off to the right start. And who knows—once they see how open you are to taking their advice, they might even become your closest collaborators and most ardent supporters.

The likelihood of you, on your own, putting together a winning strategy to achieve your vision is extremely low. This is why the world needs cooperation. The more experience that you can incorporate into your initiative, the greater your likelihood of success is going to be.

4. BUILD JOINT OWNERSHIP

One common question asked about cooperation is "How do we create buy-in without asking for money?" The answer is responsibility. The sentinel meerkat standing on the rock who keeps watch on behalf of the other meerkats has a responsibility—identify predators and warn the group. Although meerkats take turns standing sentinel, whoever is on watch is assuming ownership of the task of protecting the group. If this meerkat for some reason felt that it was not her responsibility, that someone else would pick up the slack if she decided to relax for a bit, then she would put the entire group in jeopardy.

The same is true for humans. A sense of joint ownership is essential for cooperation. People need to feel that they are critical to the success of the group. Otherwise, there is no point in their putting in the effort. I have seen lack of joint ownership be the downfall of many great initiatives. An organization will draft a strategy and then send it out for public consultation. They may get a few passionate critiques from people who could be bothered putting in a response, but they are not going to get buy-in for the initiative. Why? Because they were never really open to joint ownership from the beginning. Those who drafted the strategy wanted to own it and control it, and hence, the public consultation was really just an act of appeasement.

People are not stupid. They can tell the difference between meaningful participation and cursory consultation just to fulfill a mandate of "stakeholder engagement." People need to feel that they are critical to the success of your shared vision, not just following and commenting on it. To build a cooperative movement, you don't need more "followers"; you need owners—people who are willing to put their shoulder to the wheel. When you invite people to truly participate in the shaping of an initiative, you are giving them joint ownership—that is, the power to alter the course of the initiative. Joint ownership requires a big dose of humility, because you must be willing to have your ideas torn to pieces and replaced by something better. If you are willing to do this, you will get joint ownership. And you will get a loyal group of people who are willing to stick their necks out to ensure that your collective vision is achieved.

5. REMEMBER WE ARE ANIMALS

We know that animals have individual personalities. Sharks have shown us that different individuals of the same species and even the same population are bound to have different approaches to get to the same place. Long-lived animals like elephants spend years getting to know each other's quirks and figuring out how they should interact and communicate with each individual in their group. These animals know that there is no one-size-fits-all approach to members of their species. Why should we be any different?

One thing I love about building cooperation is that you get to understand just how eclectic people are. Some people are extremely opinionated, boldly spoken, and reluctant to be wrong. Others need to think aloud, formulating their opinions as they are speaking to the group. Others will sit there in a meeting for hours in silence and then suddenly come out with a piece of gold that completely changes the direction of the meeting. And then there are people like me who love to synthesize this discussion and find new connections that were not obvious at the start. To be successful in achieving your shared vision, you need all these individuals. You need all their quirky personalities.

You need their unique insights. Like I said in the first chapter, I see cooperative problem-solving as a multi-sided polygon. Not all of us see the problem from the same angle, so the more angles you have on that polygon, the closer you're going to get to a perfect circle.

Getting to know people on an individual level is fundamental to the early successes of a cooperative movement. This means knowing people's names, the names of their children and pets, their hobbies, and their likes and dislikes. Often it also includes seeing people in their own environment. I once drove an hour and a half up the coast to meet a veterinarian for the first time because she had expressed interest in the shared vision that we were building. We met at a café in her town. I was only a green researcher, and she could have easily questioned my authority, but she said that she was so impressed that I took the time to go and see her that she knew that we were onto something good. Establishing trust with other people, as the social animals that we are, is at the heart of cooperation.

When initiatives grow larger, there is a tendency to forgo such intimate interactions among humans for more structured processes and forms of communication. Such a switch can be tempting, given how onerous it can become to keep in touch with individuals as an initiative expands. The mass email replaces our out-of-the-blue phone conversations. The Zoom or Teams meeting replaces catching the train across the city to a cozy café, where we could sit for a few hours and share a few laughs and look each other in the eye. It is flawed to assume that over the long run, impersonal interactions will yield the same success as a series of more personal interactions, for which our brains are perfectly wired. We are social animals. We sniff each other. We mirror each other's movements. We need regular physical contact with each other to cooperate at our best.

6. GO SLOWLY AND BE FLEXIBLE

I know it seems ludicrous to suggest that we solve the world's problems slowly, especially when they are so pressing. But I have learned that the slow approach is the one that eventually gets you over the mountain. When we look at what others are doing, and especially when we compare their successes with ours, it creates a false sense of urgency. I know—I have been prodded by this false sense of urgency to accomplish the impossible in record speed for most of my life. We rush to get things established, rush to get wins on the board, and rush to show everyone that what we're doing has an impact.

Cooperation requires trust, and trust takes time. The start of it often involves people hearing about your initiative from other people whom they trust. Word of mouth has a slow rate of spread, but it pays in trust. Once they're in contact with you, people have to like you, believe in the sincerity of what you're doing, and feel valued. A little trust is built. But they need more than that—for people's trust to grow, they need a sense of belonging, transparency, and ownership. It would seem that our human brains are so well programmed for social interaction that there is no good way of short-cutting this process. We must build trust slowly, as it requires consistent positive interactions over time. When you build cooperation slowly, you bring everyone along with you on the journey. People feel included, which leads to more cooperation. Little by little, everyone's trust grows, until you have a powerful cooperative movement. You take your time in creating quality solutions that actually have an impact. And because you have grown at a human pace, you will have something of value that can stand the test of time.

Planning in a chaotic world is difficult, which is why cooperative movements need to be flexible. In the past, organizations invested heavily in long-term planning and Key Performance Indicators, but now many successful businesses use leaner forms of planning, realizing that it can be harmful to place such rigid expectations on the future. I still think that strategic planning is a useful exercise, but not for the same reasons we once believed it was. Developing a strategic plan is essential to clearly articulate a shared vision and is the easiest way to ensure that everyone is on the same page and feeling optimistic about the purpose of the initiative. But a strategic plan is only meant to serve as the foundation of a building. If you find yourselves going too granular, defining exact activities, tasks, roles, responsibilities, or timelines—turn around. Strategic plans are high level for a reason, and that is to create joint intentionality. They are not meant to dictate the future. Animals cooperate because it gives them flexibility to respond to a wide variety of environmental changes. Humans, too, need to be given latitude to use cooperation to approach any strategy in a multitude of ways as situations change.

I highly recommend that cooperative initiatives look into the business literature around design thinking and agile development. Both of these methods were created in response to a chaotic, rapidly changing environment in information technology, and they have both been successful in a variety of operational contexts. I've relied on these two methodologies with all of the cooperative initiatives in which I've been involved.

Taking a learn-as-you-go approach is often criticized for being too slow, but I have found that attempting to stick to rigid action plans in the face of changing outside circumstances is actually much slower in the end. If evolution has taught us one thing, it's that those who can successfully adapt to changes will survive. Cooperation is one of the major adaptations that have ensured humans' survival thus far, so really, we are living case studies of design thinking and agile development.

7. WHEN FACED WITH CHAOS, COME BACK TO YOUR VISION

I chose to finish this book with the story of the elephants herding together in response to the strange music that they were hearing at Melbourne Zoo because I think it provides the perfect illustration of what animals do in the face of chaos.

In the 1960s, when there were growing concerns that the U.S. government couldn't be trusted and that the world might be annihilated by nuclear war, we saw humans begin to pull away from each other. I believe that this reaction was made worse by the media coverage that people were watching on television. What we saw in the COVID-19 pandemic is that alarming media stories, like videos of people fighting over toilet paper and images of overflowing morgues, caused people all over the world to panic and behave irrationally. Humans, like many other cooperative animals, do herd together during times of uncertainty, but if we are constantly exposed to messages that the world is a scary place and people cannot be trusted, we may just decide that the safest thing to do is to go it alone.

Chaos is bound to happen. It is the way of nature. Pandemics will come. Natural disasters will occur. Human-made disasters will continue to take place and to shock us and to make us question our resolve in humanity's capacity to work together. But one of the many valuable things that other animals can teach us is that there is safety in numbers. We can pull together during these times of difficulty, these times when we're afraid.

When the Space Shuttle *Challenger* exploded in mid-air on January 28, 1986, breaking apart 73 seconds into its flight and killing all seven crew members aboard, it shook NASA and many Americans to their core. But ultimately, they put their fear aside and continued to explore space because they returned to a shared vision that was bigger than their fear. Alone, they could not have continued to pursue the crazy ambition of space travel. But together, they could. And together, we can.

The call to action is quite simple. We must make the choice to work together to improve our world rather than sit around in inaction and pessimism. We must do it today. We must do it tomorrow. And we must do it every day for the rest of our lives. It will not get easier. If anything, the further down the road we are, the harder it will get. But by cooperating, we will be surrounded by our fellow human beings who share our passion and our vision, and this is the fuel that will keep us going.

V.

I hope this exploration has shown you, as it has me, that cooperation is possible. We now have evidence to confirm what Charles Darwin suspected—that cooperation is a natural behavior found widely across the animal kingdom. It does not require sainthood or political prowess. Cooperation is the realm of ordinary humans, the realm of whales, meerkats, sheep, elephants, dogs, wolves, and birds, and the realm of you and me.

Our natural response to chaos is to cooperate. In the beginning of humanity, we did just this. We were groups of hominins living together, sharing our resources with each other, walking across Africa together, picking each other up when someone fell down, taking care of one another when someone fell sick or injured. That was all that humanity was then—a group of people cooperating. And I would argue that this is still what humanity can be now. The International Space Station has shown this, the Egypt–Israel Peace Treaty has shown this, the businesses in 1% for the Planet have shown this, and the community of Cabo Pulmo has shown this. These are all real people like you and me, tackling real-world challenges by working together. They are testament to the reality that humans have not lost our cooperative nature. We simply forget what our species is capable of, what our species was built for: cooperating through chaos.

When I was living in Miami, the Quakers taught me an old saying, attributed to William Penn: "Let us see what love can do." When I first heard it, I was floored by its simplicity. Since then, those words have resonated in my mind.

In concluding this book on cooperation, I'm not saying that cooperation will definitely solve all of the world's problems—there is no way to guarantee that. I'm simply saying this:

Let us see what cooperation can do.

LET ME KNOW!

If you enjoyed this book, I need something from you please: I need you to review the book. I'm depending on your cooperation.

Writing a book is hard work. It took me 4 years and dozens of drafts to write this book, which involved researching and making sense of over 500 references. This book tested me mentally. During the 4 years, my family, friends, and body paid the price of the hours I spent at my desk and the stress I underwent.

Despite the challenge of it, I love writing. I'd like to keep writing books like this one until I die, but I need to know that my books are making an impact in the world. A desire for impact is at the core of my being.

What this means is that I need you to please leave a review if this book made any impression on you.

Good or bad, let me know what you honestly thought of it. Note that Amazon doesn't allow reviews from family or friends. If we're not family or friends, please leave a review on Amazon, Barnes & Noble, or the website where you bought the book. If you know me personally, please leave a review on Goodreads. It's easy to create an account, and lots of readers like myself trust Goodreads. If you bought the book at a bricks-and-mortar independent bookshop (go you!), please call them and tell them you enjoyed the book so that they know to order more.

I read every single review myself, and I use the feedback from reviews to improve my writing.

If I receive your review, I'll know my book has made a difference, which will signal to me that I should write further books. I'll be forever grateful to those who choose to help me on my writing journey.

THANKING YOU IN ADVANCE FOR YOUR ACT OF COOPERATION

CHAPTER 12 REFERENCES

Free Resources

Learn more about how elephants communicate with one another, and support ongoing research and conservation efforts at ElephantVoices: https://www.elephantvoices.org/

Further Reading

de Waal, F. (2016). *Are we smart enough to know how smart animals are?* W.W. Norton & Co. https://wwnorton.com/books/Are-We-Smart-Enough-to-Know-How-Smart-Animals-Are/

Goodall, J. (2000). *Reason for hope: A spiritual journey.* Grand Central Publishing. https://www.hachettebookgroup.com/titles/jane-goodall/reason-for-hope/9780446676137/?lens=grand-central-publishing

Lamott, A. (2014). *Bird by bird: Some instructions on writing and life.* Knopf US. https://www.penguin.com.au/books/bird-by-bird-9780385480017

Pressfield, S. (2013). *The war of art.* Black Irish Entertainment LLC. https://blackirishbooks.com/product/the-war-of-art/

References

1. Garcia, M. A. (2023, October 2). *International Space Station.* NASA. https://www.nasa.gov/reference/international-space-station/

2. Hall, A. (2023, September 27). *Station facts.* NASA. https://www.nasa.gov/international-space-station/space-station-facts-and-figures/

3. Evans, C. A., & Robinson, J. A. (n.d.). *Space station orbit tutorial.* Gateway to Astronaut Photography of Earth. Retrieved March 22, 2024, from https://eol.jsc.nasa.gov/Tools/orbitTutorial.htm

4. Dobrijevic, D. (2023, November 13). *Track the ISS: How and where to see it.* Space. https://www.space.com/how-to-track-the-international-space-station

5. Hawkins, T. (2023, July 17). *60 years and counting: President Eisenhower signs NASA into existence.* NASA. https://www.nasa.gov/history/60-years-and-counting-president-eisenhower-signs-nasa-into-existence/

6. Brazda, T. (2023, August 23). *Space station 20th: Historical origins of ISS.* NASA. https://www.nasa.gov/history/space-station-20th-historical-origins-of-iss/

7. Howell, E. (2024, February 24). *International Space Station: Everything you need to know about the orbital laboratory.* Space. https://www.space.com/16748-international-space-station.html

8. Hazaimeh, H. (2020, April 16). *When Sadat went to Israel.* Arab News. https://www.arabnews.com/node/1659751

9. Regev, M. (2022, November 24). *Looking back at Egypt's Anwar Sadat's historic Jerusalem visit, 45 years ago - opinion*. The Jerusalem Post. https://www.jpost.com/opinion/article-723321

10. Begin, M. (1977, November 11). *Address By Prime Minister Menachem Begin to the Egyptian people*. https://db.begincenter.org.il/en/article/address-by-prime-minister-menachem-begin-to-the-egyptian-people/

11. 1% for the Planet. (2023). *Our Story*. https://www.onepercentfortheplanet.org/about/story

12. 1% for the Planet. (2023). *Membership*. https://www.onepercentfortheplanet.org/membership

13. Aziz, A. (2023, May 9). *How businesses and individuals are leading the way with 1% for the Planet in environmental giving*. Forbes. https://www.forbes.com/sites/afdhelaziz/2023/05/09/how-businesses-and-individuals-are-leading-the-way-with-1-for-the-planet-in-environmental-giving/

14. Earth911. (2022, December 12). *Why should you care about 1% for the Planet?* Earth Media Partners. https://earth911.com/business-policy/one-percent-for-the-planet/

15. 1% for the Planet. (2022). *Annual Report 2021*. https://assets-global.website-files.com/634735ad1c3cc168fc2117a1/63e54937dd05b749d65dc994_2021%20One%20Percent%20for%20the%20Planet%20Annual%20Report.pdf

16. Solar Sister. (2018). *Our Partners*. https://solarsister.org/our-partners/

17. Kenny, S. (2022, December 1). *A deep dive into Mexico's Sea of Cortez, the world's aquarium*. Much Better Adventures. https://www.muchbetteradventures.com/magazine/mexico-sea-of-cortez/

18. UNESCO. (2024). *Islands and Protected Areas of the Gulf of California*. https://whc.unesco.org/en/list/1182/

19. Radwin, M. (2021, October 30). *Local communities saved Cabo Pulmo with a national park. Then came the tourists*. Mongabay. https://news.mongabay.com/2021/10/local-communities-saved-cabo-pulmo-with-a-national-park-then-came-the-tourists/

20. Folger, T. (2012, December 17). *Saving the wonders of the Sea of Cortez*. OnEarth Magazine. https://mission-blue.org/2012/12/saving-the-wonders-of-the-sea-of-cortez/

21. Albritton, A. (2024). *Cabo Pulmo National Park: A community story*. PADI. https://blog.padi.com/cabo-pulmo-national-park-a-community-story/

22. Smithsonian National Museum of Natural History. (2018, April). *Cabo Pulmo Protected Area*. Ocean. https://ocean.si.edu/conservation/solutions-success-stories/cabo-pulmo-protected-area

23. WWF International. (2019, June 17). *The story of Cabo Pulmo*. YouTube. https://www.youtube.com/watch?v=Rk-qLXek7yQ

24. Poole, J. H., & Moss, C. J. (2008). Elephant sociality and complexity: The scientific evidence. In C. Wemmer & C. A. Christen (Eds.), *Elephants and ethics: Toward a morality of coexistence* (pp. 69–98). John Hopkins University Press. https://www.press.jhu.edu/books/title/9599/elephants-and-ethics

25. McComb, K., Moss, C., Sayialel, S., & Baker, L. (2000). Unusually extensive networks of vocal recognition in African elephants. *Animal Behaviour, 59*(6), 1103–1109. https://doi.org/10.1006/anbe.2000.1406

26. Archie, E. A., Moss, C. J., & Alberts, S. C. (2006). The ties that bind: Genetic relatedness predicts the fission and fusion of social groups in wild African elephants. *Proceedings of the Royal Society B: Biological Sciences, 273*(1586), 513–522. https://doi.org/10.1098/rspb.2005.3361

27. Luntz, S. (2019, November 9). *A zoo is providing live concerts to entertain its elephants.* IFLScience. https://www.iflscience.com/a-zoo-is-providing-live-concerts-to-entertain-its-elephants-54172

28. 7News Melbourne. (2019, May 28). *New additions to Melbourne Zoo's elephant enclosure.* Facebook. https://www.facebook.com/watch/?v=348339979216755

29. Plotnik, J. M., Lair, R., Suphachoksahakun, W., & de Waal, F. B. M. (2011). Elephants know when they need a helping trunk in a cooperative task. *Proceedings of the National Academy of Sciences, 108*(12), 5116–5121. https://doi.org/10.1073/pnas.1101765108

30. Plotnik, J. M., & de Waal, F. B. M. (2014). Asian elephants (*Elephas maximus*) reassure others in distress. *PeerJ, 2*, e278. https://doi.org/10.7717/peerj.278

31. Payne, K. B., Langbauer, W. R., & Thomas, E. M. (1986). Infrasonic calls of the Asian elephant (*Elephas maximus*). *Behavioral Ecology and Sociobiology, 18*(4), 297–301. https://doi.org/10.1007/BF00300007

IT TAKES A VILLAGE TO RAISE A BOOK

To Keyoren, for believing in me and lovingly supporting me through all the ups and downs of writing this challenging book; to my parents, Rick and Sue, and my brother, Luke, for telling me I was a good writer long before I believed it; to Mrs. Hitman, for teaching me how to write well and encouraging my writing career from the time I was in high-school English; to Kat Jones, for igniting my interest in biology; to Sarah, for overanalyzing human behavior with me and making me burst into laughter during even my darkest times; to Ariel and Aubrey, for accepting me for who I am; to Julie Frentzel, for showing me how to enjoy life; to Grandma Harriet, for encouraging me to see the world; to my Aunt Jo, for fostering my love of science; to Natalie Goldberg and Julia Cameron, for showing me how to walk down the writing path; to the Leonard and Jayne Abess Center for Ecosystem Science and Policy, for molding me into an interdisciplinary scientist and a big-picture thinker; to my Abess Center mentors—John Beier, Gina Maranto, and Kenny Broad—for embracing and honing all my wild ideas; to Bob Fagen, for pointing me in the right direction in the vast sea of literature on cooperation; to Bruce and Sheryl Sagnol, for teaching us to sail and to keep sailing even when we're afraid, and to Bruce for putting me onto the SOFAR channel; to Alan, Dom, and Shane, for introducing me to animal welfare science; to Elise, for cheering me on always and supporting me to become a consultant; to Nick, for drawing me in to the fascinating zoo world; to Celine, Sylvester, Ray, Kaz, Meyrelle, Andrew, and Ruby, for making me part of the family and showing support of my writing and our sailing adventures; to my talented editor, Tina Morganella, for wading through my mess of American-Australian spellings and making me a better writer through positive reinforcement; to Julia at Le Bon Choix, for asking about my book with each matcha latte I ordered and eagerly awaiting its release...

And to floppy-eared Leo, who, through rewrite after rewrite, always stayed by my side.

Thank you.

STAY IN TOUCH...

Thank you, dear reader, for giving me your time. If you'd like to collaborate with me in shaping a better world for animals (yes, please!), provide feedback on this book, or share information that you think would be of interest to me, please send me a message at www.jillfernandes.com. I may need some time, but I will respond. :)

Until next time,

Jill

* * *

ABOUT THE AUTHOR

Dr. Jill Fernandes is an animal welfare scientist and consultant dedicated to shaping a better world for animals. Growing up on a farm in Missouri surrounded by cattle, horses, dogs, cats, foxes, deer, and quail, Jill has a deep-seated interest in the workings of nature and the social behavior of animals, including humans.

Jill has a PhD in Environmental Science and Policy from the Leonard and Jayne Abess Center at the University of Miami. She is a Fulbright Scholar and a Graduate Research Fellow of the National Science Foundation. She undertook interdisciplinary research on a range of topics in the environmental and animal sciences at The University of Queensland before moving into animal welfare consulting. Through her company, Animal Centric, she writes about current science on animals, offers practical courses, and develops strategies with zoos, aquariums, farms, and other organizations with animals under their care to ensure that all animals can live the best lives possible.

Jill's intent in writing *Cooperating Through Chaos* is that the book be a catalyst for cooperation everywhere.

WWW.JILLFERNANDES.COM

Made in United States
Troutdale, OR
10/27/2026

56543062R00215